Secrets
of 5-Axis
Machining

by Karlo Apro

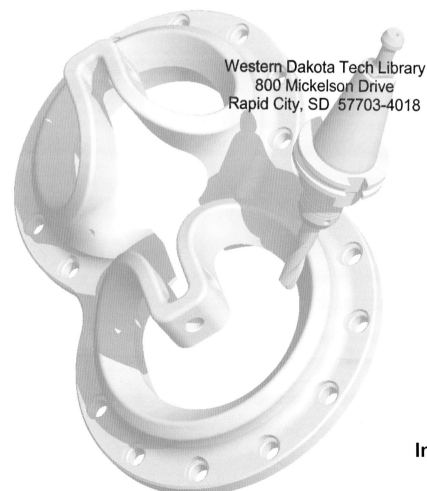

Industrial Press, Inc.
New York

Library of Congress Cataloging-in-Publication Data
Apro, Karlo.
  Secrets of 5-Axis Machining / Karlo Apro.
     p. cm.
  Includes index.
  ISBN 978-0-8311-3375-7
  1. Machine tools--Numerical control. 2. Machining. I. Title. II. Title:
Secrets of 5-Axis Machining.
  TJ1189.A68 2008
  671.3'5--dc22

                                     2008027258

Industrial Press, Inc.
989 Avenue of the Americas
New York, NY 10018

First Printing, August, 2008

Sponsoring Editor: John Carleo
Interior Text and Cover Design: Paula Apro
Developmental Editor: Robert E. Green
Production Manager: Janet Romano

                10   9   8   7   6   5   4   3   2   1

Printed by Thomson Press India Limited

Dedication

This book is dedicated, in loving memory, to my mother Piroska. She taught me the meaning of hard work and perseverance. Although she passed away before the completion of this book, her spirit continues to live with me.

Acknowledgements

I would like to thank Yavuz Murtezaoglu for giving me the inspiration to write this book.

A special thanks to Laura Norton for her humbling insights.

And above all, I would like to thank Paula Apro, my hard-working wife, friend, editor, designer, and manager. For without her this book would never have come to be.

All the images in this book, including the virtual machines, were modeled using Mastercam® (CNC Software, Inc.). The virtual machines were brought to life using the machine simulation capabilities of MachSim (ModuleWorks) and VERICUT® (CGTech).

For more information on these products or companies please contact:

CNC Software/Mastercam	*MachSim/ModuleWorks*	*CGTech/VERICUT*
671 Old Post Road	*ModuleWorks GmbH*	*9000 Research Drive*
Tolland, CT 06084	*Ritterstr. 12 a*	*Irvine, California 92618*
860.875.5006	*52072 Aachen, Germany*	*949.753.1050*
www.mastercam.com	*+49.241.4006020*	*www.cgtech.com*
	www.moduleworks.com	

For more information on the author, please visit www.multiaxissolutions.com.

Table of Contents

Introduction **1**

Chapter 1: History of 5-Axis Machines **3**

Common Misconceptions 4

Reasons to Use Multiaxis Machines 8

Chapter 2: Know Your Machine **13**

Multiaxis Machine Configurations 14

Table/Table Multiaxis Milling Machines 18

Machine Rotary Zero Position (MRZP) 21

Nesting Positions. 26

Rotary Table Dynamic Fixture Offset 27

Head/Table Multiaxis Milling Machines 31

Head/Head Multiaxis Milling Machines 36

Finding the Pivot Distance 37

4–Axis Machines 39

General Maintenance & Issues for Multiaxis Machines . . . 40

Milling Machines With Five- or More-Axes. 43

Chapter 3: Cutting Strategies **45**

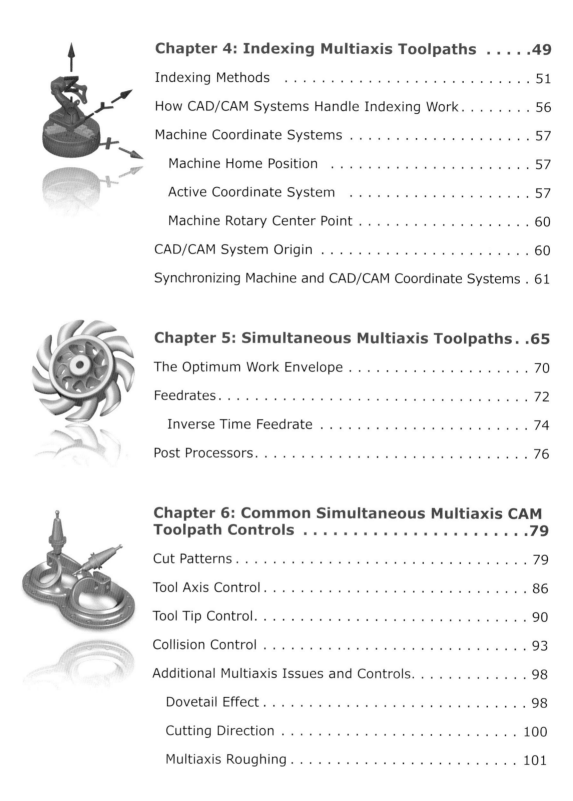

Chapter 4: Indexing Multiaxis Toolpaths 49

Indexing Methods 51

How CAD/CAM Systems Handle Indexing Work 56

Machine Coordinate Systems 57

    Machine Home Position 57

    Active Coordinate System 57

    Machine Rotary Center Point 60

CAD/CAM System Origin 60

Synchronizing Machine and CAD/CAM Coordinate Systems . 61

Chapter 5: Simultaneous Multiaxis Toolpaths . .65

The Optimum Work Envelope 70

Feedrates 72

    Inverse Time Feedrate 74

Post Processors 76

Chapter 6: Common Simultaneous Multiaxis CAM Toolpath Controls 79

Cut Patterns 79

Tool Axis Control 86

Tool Tip Control 90

Collision Control 93

Additional Multiaxis Issues and Controls 98

    Dovetail Effect 98

    Cutting Direction 100

    Multiaxis Roughing 101

Chapter 7: Machine Simulation 103

G-code Simulation Versus CAM Simulation 105

Configuring Virtual Machines For Simulation 105

Virtual Machine Building. 106

    The Skeleton 106

    Components vs. Models 107

Machine Simulation Interfaces 116

Using Machine Simulation 117

Chapter 8: Selecting The Right Machine For Your Application 119

Head/Head Machines *(with long X or Y – axis linear travel, but limited rotary axes travel)* 121

Head/Table Machines *(with long X-axis travel)* 123

Head/Table Machines 126

Rotary Table – Tilting Head Combinations 128

Table/Table Machines. 132

Gantry Type Head/Head Machines. 134

Chapter 9: Choosing a CAD/CAM System For Your Application. 137

Special Purpose Software. 137

CAD/CAM Toolbox 139

Multiaxis CAD/CAM Considerations 139

Multiaxis CAM. 140

Multiaxis CAD/CAM Training 144

Behind the Scenes: CAD/CAM Software Development . . 145

General Guidelines for Researching CAD/CAM Software. . 146

Chapter 10: Putting It All Together 149

Why Use Multiaxis Machining Techniques? 152

What is a Standard 5-Axis Machine? 153

What is the Standard Axis Convention? 154

What are the Three Major Multiaxis Machine Types? . . . 154

What are the Major Building Blocks of a CNC Machine? . 156

What are the Most Important Physical Positions of a
Multiaxis Machine? 157

What Tools are Needed to Find MRZP? 159

Description of Indexing/Rotary Positioning Work 159

What is a Post Proccessor? 159

Definition of an Axis 160

Defining a Simultaneous 5-axis Toolpath 160

What are the Three Common Simultaneous Multiaxis CAM
Toolpath Controls 161

Multiaxis Machine Offsets 161

Finding Machine Rotary Zero Position 162

Finding the Pivot Distance 164

Indexing/Rotary Position Work Overview 166

Picking a CAD/CAM System for Multiaxis Work 166

Machine Simulation 167

Conclusion 167

Introduction

Are you utilizing 5-axis machining? Could your shop benefit from the efficiency and power that 5-axis machining offers? The majority of people not embracing this technology lack a true understanding of 5-axis practices. There are many common misconceptions on the subject, and the intent of this book is to demystify 5-axis machining and bring it within the reach of anyone interested in using the technology to its full potential. The information presented in this book was gathered during 30 years of hands-on experience in the metal-working manufacturing industry — bridging countries, continents, and multiple languages (both human and G-code.) The author worked in Hungary, Germany, Canada, and the USA, specializing in multiaxis solutions. He spent many years setting up, programming, and repairing CNC equipment, and has used a number of different CAD/CAM systems. He has worked as a self-employed multiaxis consultant, as well as directly for CGTech (the makers of VERICUT®) and CNC Software Inc. (the makers of Mastercam®.)

The author has instructed countless multiaxis training classes over the past decade. These classes covered topics such as operating CNC equipment, programming CNC equipment, both manually and with CAD/CAM systems, and building virtual machines with different verification systems. Through the years, the author has met many professionals around the world and has come to a realization that they all have the same questions, misconceptions, and concerns, when it comes to 5-axis machining. The need for unbiased information on the subject became apparent.

Up to this point, the best way to get information on 5-axis machining was to talk to peers in the industry, in the hope that they would share what they had learned. Visiting industrial trade shows and talking to machine tool and CAD/CAM vendors are other options — except that these people all give their individual points of view and will promote their own machine or solution. Everybody claims to have the best mouse-trap, and it is left to the individual to choose the right one.

This book is not a training manual for any particular machine or CAD/CAM system. Rather, it is an overview of multiaxis machine types and the common control methods that CAD/CAM systems use to drive the machines. The book will guide you through this realm, from basic to complex concepts, and will provide information to help you choose the right tools, including the machine, work-holding method, CAD/CAM system, and machine simulation package that will best suit your specific application. The book contains numerous illustrations to help you to precisely implement these tools.

History of 5-Axis Machines

Long before CNC controllers appeared, 4-5-6-12- and more-axis machines, referred to as multiaxis machines, were being used. The individual axes were controlled mechanically through levers riding on cam plates. Some machines had more than 12 cam plates, controlling not only tool/table and rotary motions, but also clamping and unclamping of work-holding fixtures. These machines were cumbersome and time consuming to set up, but they were perfectly suited for mass production.

The first NC (numerical control without internal memory) machines were cumbersome to set up and operate, but they also were great for mass production. At first, only the most affluent and established shops could afford them. Programming was a lengthy, error-prone process. Soon, machine builders added internal memory to their controllers, then they added the ability to execute simple branching looping logic, and to call subroutines from other subroutines. It was possible to use these macro languages directly on the machine and to quickly change set-ups, especially for family type parts. Different machine builders developed various solutions, which created a number of CNC (computer numerical control with internal memory) programming languages. Companies with familiar names like Fanuc, Acramatic, Heidenhein, Siemens, Mazatrol, etc., all developed their own languages, but these quickly became an issue. Some shops ran ten machines with eight different languages. If a repeat job came in, and the originally programmed machine was busy, a new program would have to be re-written from scratch because of the language differences.

Next, the first rudimentary CAD (Computer Aided Design)/CAM (Computer Aided Manufacturing) systems were developed. At first, these software solutions were introduced by the same companies that developed the controllers. Soon after, enterprising individuals wrote their own CAD/CAM software. This jump in technology was huge because it allowed engineers to draw their parts in a CAD program, generate a toolpath in the CAM system's generic language, and then translate it into multiple G-Code languages quickly, using the appropriate post processor.

This progress meant that CNC machines were no longer the exception, and they started to become the norm. They were no longer used only for mass-production and they became versatile, accurate, and affordable.

Multiaxis machines went through a similar process, but because they were more complicated, this process took longer. First, the machines were expensive to purchase and maintain, and harder to program. Only large aerospace companies had the need, the money, and the personnel to handle multiaxis applications. Some companies kept their own processes closely guarded in order to gain an advantage. Many software packages were born out of necessity — in order to solve specific application challenges. Software, in general, is always on the very leading edge of technology — pushing the limits of software possibilities and hardware restrictions.

Today, there are many machine builders offering a variety of multiaxis equipment in a wide range of configurations, quality, and price. Computers have become very affordable, and CAD/CAM systems now offer excellent multiaxis cutting strategies with great tool control and large post-processor libraries. As a result, even smaller shops can, and do, implement multiaxis machining.

Most machine builders are expanding production and embracing new technology. Many believe that it is imperative to compete in the global market, especially against countries with abundant cheap labor. This attitude has resulted in increased multiaxis machine sales and some machine builders now have waiting lists of customers for multiaxis machines. Multiaxis machining is a constantly expanding field, with almost endless possibilities.

Common Misconceptions

Most people associate the word "5-axis" with complicated motions such as those for the induction pump illustrated in Figures 1-1 and 1-2, and the programming techniques needed. This view is reinforced by visits to any industrial trade show to see both machine builders and CAD/CAM vendors showing off their most complicated creations.

Figures 1-1 *Example of induction pump set-up*

Figure 1-2 *Example of induction pump design.*

In reality, the majority of 5-axis users don't ever make an impeller, or finish ports for a racing-engine cylinder head. Most of them machine parts using simple 3-axis drilling, contouring, and pocket milling routines, while rotating the part occasionally in a rotary indexing mechanism, as illustrated in Figures 1-3 and 1-4. Very elaborate parts can also be machined by applying 3D surfacing toolpaths and engaging the part from different angles by indexing a rotary table.

Figures 1-3 and 1-4 *Examples of positioning work.*

Using a multiaxis machine will greatly simplify the motions required, the programming effort, and the amount of fixturing needed for machining complex workpieces. Other benefits include the elimination of multiple set-ups, increased accuracy, and better surface finish.

Many shops are currently making parts by moving them manually to different fixtures on 3-axis machines. Compared with this procedure, production can be increased greatly without much effort by using a 4- or a 5-axis machine. If simply a single- or dual-rotary indexing table was added, only slight edits would be needed to the CNC-code files. Examples are shown in Figures 1-5 and 1-6.

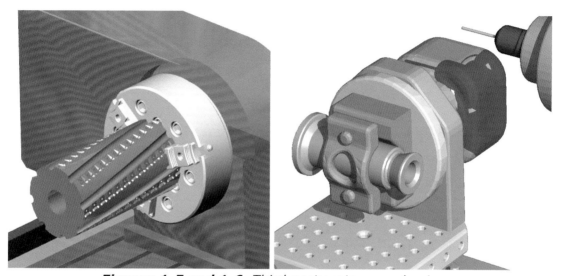

Figures 1-5 and 1-6 *Third-party rotary mechanisms.*

Moving to multiaxis machining requires thinking in space instead of in a flat plane. Dedicated multiaxis machines have been developed for the kind of indexing work shown in the accompanying Figures 1-7 and 1-8, using tombstone type fixtures.

Figure 1-7 *Example of tombstone fixture.*

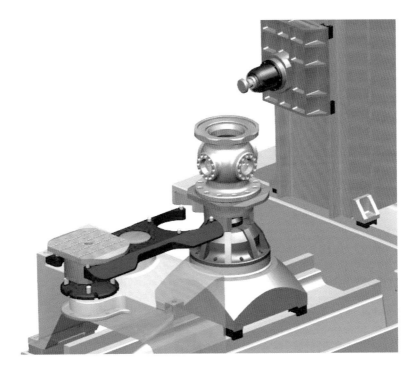

Figure 1-8 *Example of 4-axis positioning.*

Once you enter the multiaxis realm, new doors will be opened for your shop. Your company will quickly become more adept and able to tackle more complex work. Before too long, your shop will start taking on more and more jobs, and will need to be expanded.

> **Common Misconception: 5-axis CAD/CAM is too expensive and is hard to use.**

The above statements were true in the past, but not any more. If you currently own a CAD/CAM system, there is a good chance you already have 5-axis positioning capabilities. Most CAD/CAM systems include these capabilities in their base package. Many times, it is just a matter of training that is needed to get up and running.

When you are shopping for a CAD/CAM system, make sure to choose one from a reputable company with a commitment to training and local support. Remember that a CAD/CAM system is just another tool in your tool belt. You can buy fancy tools that are very capable, but they are worthless if you don't know how to use them. Great local support may very well be the most important feature of your new tool.

If you do a lot of simultaneous multiaxis work, the price of the CAD/CAM will be only a small factor. More training will be needed, but you will be able to charge almost double for your hourly machine time. The 'hard to use' part always comes down to training — was it easy to learn how to operate your first CNC machine?

Don't enter the multiaxis world by starting with a complex, simultaneous job. If you already own a 3-axis machine, start with a single- or dual-rotary table and apply indexing techniques. You will make parts faster and more accurately, and you will be able to invest in more equipment. When you decide to buy new equipment, see if you can bundle a CAD/CAM purchase with the machine's purchase order. This is also a good time to make sure your CAD/CAM system speaks your specific machine's language — in other words, that it has the correct post processor.

Some companies buy equipment with a turn-key solution, which ensures that their specific job will run on the machine upon delivery from the manufacturer. Many machine tool builders employ capable teams of applications engineers, who in turn, work closely with CAD/CAM developers. Together, the teams determine the most efficient way to machine any specific part, based on many factors such as; material, quantity, tolerance requirements, and tooling availability.

Reasons to Use Multiaxis Machines

Reduced Set Up Work

One important reason to use multiaxis machines is to reduce set-up time for parts such as those shown in Figures 1-9 and 1-10. Extra custom fixturing for secondary operations is very costly and time-consuming. Most parts can be manufactured in one or two set-ups, eliminating the need for extra fixturing and time.

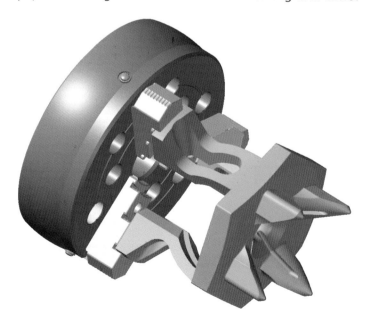

Figure 1-9 *Example part requiring positioning multiaxis machining.*

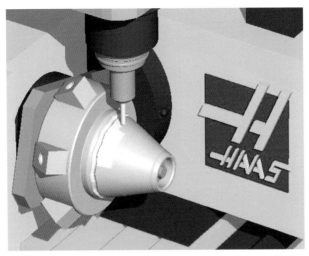

Figure 1-10 *Part requires two separate set-ups for machining.*

Accuracy

Every time you move a workpiece from one fixture to another, there is a risk of misalignment — either during the set-up itself or during operation. It is easy to build up (stacked) errors between machined surfaces when they are milled in multiple set-ups. The use of indexing rotary tables, or dedicated multiaxis machines, as shown in Figures 1-11 and 1-12, allows precise movement of short, rigid, high speed cutters for the best cutting engagement. More aggressive cuts can then be taken, with higher RPM and feed rates, while the highest levels of accuracy are maintained.

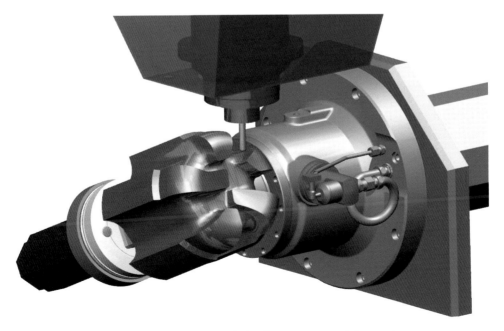

Figure 1-11 *Dedicated dual-rotary machine set-up.*

Figure 1-12 *Dedicated dual-rotary machine set-up.*

Better Surface Finishes

Using shorter tools will cause less tool deflection, which will minimize vibration and produce smooth, precise, cuts. When using ball-nose cutters it is recommended that the contact point be moved away from the tip of the cutter that isn't spinning. By tilting the tool, as shown in Figures 1-13 and 1-14, the workpiece can be engaged by a desired cutter area, which will not only improve the surface finish and repeatability, but will also greatly improve tool life.

Figures 1-13 and 1-14 *Machining parts such as these requires simultaneous cutting motions.*

Open New Possibilities

Some parts are impossible to cut on a 3-axis machine. Other parts would take too many set-ups on a 3-axis machine to be profitable. Once your shop gets comfortable with indexing work, you will be able to start machining parts such as those in Figures 1-15, 1-16, and 1-17, using simultaneous multiaxis motions, and open your business to many new possibilities.

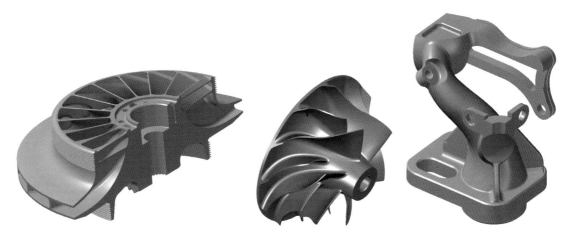

Figures 1-15, 1-16, and 1-17 *More examples of parts that require simultaneous cutting motions.*

A word of caution: Simultaneous multiaxis work is inevitably less accurate than indexing work because the machine must be run in a loose mode with the rotary drives unlocked. It is recommended that all possible roughing operations be done by indexing the rotaries to optimum angles, because the machine in locked mode is much more rigid. This type of work is also called 2+3 machining. The two rotary axes are first positioned and locked into the optimum attack position, then a standard 3-axis program is executed.

NOTES:

_____

Know Your Machine

What do you picture when you see the words "standard 5-axis machine?" Many industry buzzwords are used when describing 5-axis machines. Some of them include: staggered guide-ways, constant dynamic control, digital AC servo motors with pre-tensioned ball-screws, permanent positioning monitoring system, maximum utilization layout, long-term accuracy, and so on. To simplify things, we will say that there are three major building blocks to these types of machines.

1   The physical properties of the machine
The physical properties of the machine describe the way the axes are stacked, the rigidity and flexibility of the iron, the horsepower, torque, and maximum RPM of the spindle motor, the quality and workmanship of the guides/slides, and the rotary bearings.

2   The CNC drive system
The drive system is the muscles or the components that make the machine slides and spindles move. The system includes the servo motors, drive system, ball screws, the way positioning is controlled and monitored, and the rapid-traverse and feed capabilities.

3   CNC controller capabilities
The controller is the brain of the machine. Data handling, available on-board memory size, and dynamic rotary synchronization controls, are some of the things controlled here.

The perfect combination of the above characteristics will build a fast, accurate, easy-to-program and operate, 5-axis CNC milling machine. Many manufacturers have spent many years trying to come up with the perfect combination, and as a result there are many variations and solutions.

The illustrations in Figure 2-1 show some of the variety that exists in the machines that make up the CNC manufacturing industry.

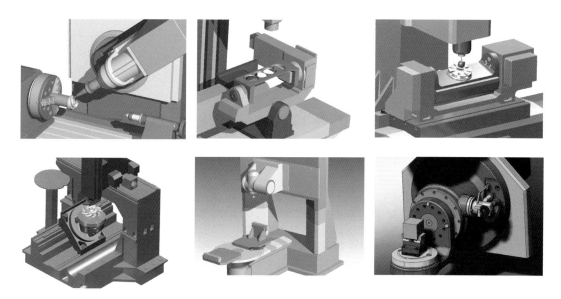

Figure 2-1 *Typical arrangements of multiaxis CNC machines.*

Multiaxis Machine Configurations

The arrangements shown in Figure 2-1 are all very popular configurations, but none of them is "standard." There is no such thing as a standard 5-axis machine. First, let's establish the definition of an axis. Any motion controlled by the NC controller, either linear or rotational is considered an axis. For instance, in the illustration in Figure 2-2, both the spindle head and the quill are capable of moving in the same direction, but are controlled by two separate commands. Movements of the head are controlled by Z and those of the quill by W.

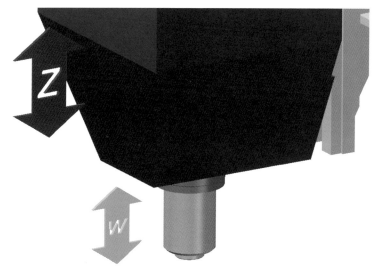

Figure 2-2 *The spindle head and the spindle quill move along parallel axes.*

The terms multiaxis and 5-axis are often used interchangeably and these terms can be confusing. The widely recognized term in the industry is 5-axis, but it is misleading because 9-axis standard possibilities exist – without adding additional sub-systems. In addition, a 4-axis machine is also considered to be a multiaxis machine. Despite the title of this book, the more accurate term multiaxis will often be used.

The following list provides the industry standard nomenclature for the basic 9-axis designations and directions.

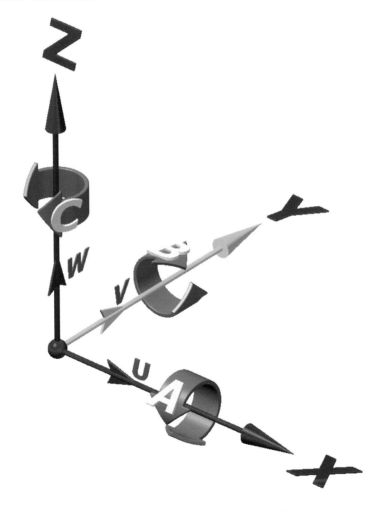

XYZ are linear axes where Z is aligned with the spindle of the machine.

ABC are rotary axes rotating around **XYZ** respectively.

UVW are parallel linear axes along **XYZ** respectively.

Unfortunately, different machine builders abide by this standard in different ways. Some builders allow the end user to change the machine's rotational directions or behavior on the fly. Third-party rotary devices, as shown in Figure 2-3 and elsewhere, can be purchased and mounted on a machine in a variety of ways. The end result of this flexibility can cause two machines, of the same make and model, to have completely different 5-axis behavior.

Every machine is a compromise of some sort. Rotational directions, start positions, and limits, will be different from one machine to another. The effective work envelope is greatly modified by changing those variables. Some rotary axes can rotate in both directions. Some axes will choose the rotary direction based on the existing position — shortest distance versus clockwise (CW) or counter-clockwise (CCW). Some machines that are equipped with **dynamic rotary fixture offset** mode will move the linear axis while rotating the rotary one based on a rotary command.

To understand these machines completely, it is necessary to look at every machine as a unique entity, to look under the skin and understand how the skeleton is constructed. You need to know where all the joints are, where the rotary axes are, where the rotary zero positions are, what makes them move, and how the whole unit functions in unison.

Different manufacturers and CAD/CAM systems have many different names for the same things. Let's establish some common terms that will be used in this book in order to avoid assumptions and confusion.

Machine Home Position (MHP) - Most machinists recognize the home position as the place to which all the axes move when you initially turn the machine on and select **Zero** return.

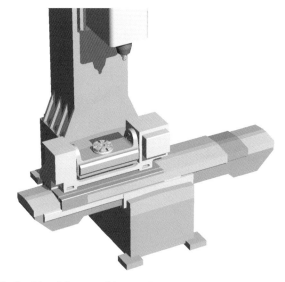

Figure 2-3 *Machine at Home Position X0. Y0. Z0. A0. B0.*

Machine Rotary Zero Position (MRZP) - On multiaxis machines, machine rotary zero shown in Figure 2-4, is at the intersection of the rotary/pivoting axes. This point may be unreachable by the machine.

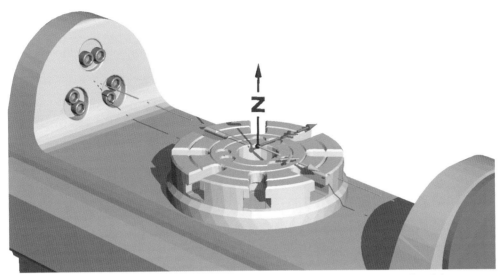

Figure 2-4 *Close-up showing Machine Rotary Zero Position.*

Program Zero Position (PZP) - Program Zero Position is the part datum in the CAM system.

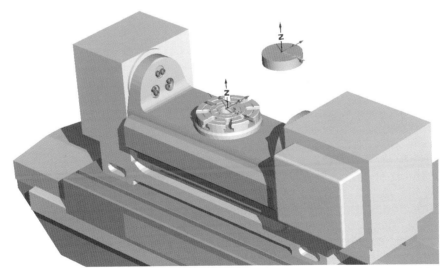

Figure 2-5 *Another view showing the relationship between Machine Rotary Zero Position and Program Zero Position.*

When setting up, operating, and programming multiaxis machines it is essential to maintain the proper relationship between the machine zero position (**MRZP**) and the program zero position (**PZP**).

If the machine does not have special features then the **PZP** must coincide with the **MRZP**.

Multiaxis milling machines can be organized further into 3 major machine types:

> **Table/Table** multiaxis machines execute the rotary motions by the dual rotary table. The primary rotary table carries the secondary rotary table, which in turn carries the fixture and the part.
>
> **Head/Table** multiaxis machines execute the rotary motions by the table, which carries the work piece. The spindle head articulates the tool with tilting motions.
>
> **Head/Head** multiaxis machines execute all rotary/pivoting motions by articulating the spindle head of the machine. The work piece is stationary.

Keep in mind that the focus of this book is milling, although the line between the mill and the lathe is blurring more and more every year. There is a new breed of multi-tasking machines available that can do milling and turning, and those are called Mill/Turn machines.

For the sake of simplicity, we will focus only on multiaxis milling machines.

Table/Table Multiaxis Milling Machines

Table/Table multiaxis milling machines can be vertical or horizontal. All the rotary motions except the spindle are done by the tables of these machines. The main rotary table carries a second rotary table, as shown in Figure 2-6, to which is fastened the fixture and the part to be machined.

Tool length offsets work the same way here as with any conventional 3-axis machine. The tool length can be changed without the need to re-post the NC data.

On these machines, the part is physically rotated around the tool. The machine's rotary devices need to be capable of handling the weight of the part and the fixture, and this capability is an important factor when rapid movements are considered. Another variation is seen in Figure 2-7.

The examples shown represent only a small fraction of the available **Table/Table** variations. Most of these machines have minimum and maximum rotary limits on one of the rotary axes. Some will have unlimited rotary motion on the other axis. Some even have the capability to spin the work, as a lathe would.

Table/Table machines are the most common types of multiaxis machines. Most people will enter the 5-axis world by purchasing a single- or dual-rotary device and bolt it to their 3-axis milling machine.

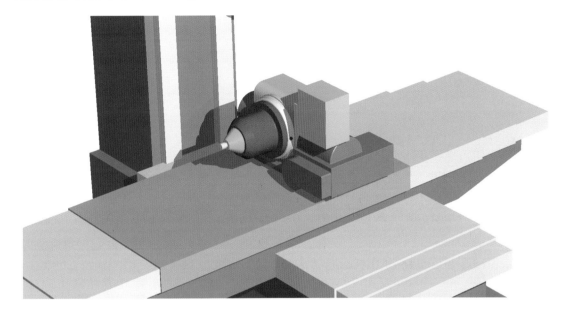

Figure 2-6 *Simulation of a dual rotary mechanism fastened to the table of a standard 3-axis CNC milling machine.*

Figure 2-7 *A third-party rotary mechanism fastened to the table of a standard 3-axis CNC milling machine.*

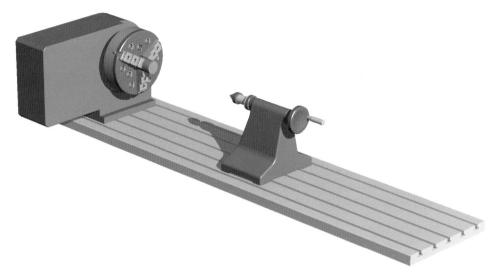

Figure 2-8 *Third-party single rotary mechanism and tailstock, fastened to the table of a standard 3-axis CNC milling machine.*

After machining one side of the work piece it is possible to index the rotary unit to machine the second side, and so on. This type of work is called indexing or positioning work. Some manufacturers use specialized dual rotary mechanisms, such as the one shown in Figure 2-9, which is designed for machining internal combustion engine components.

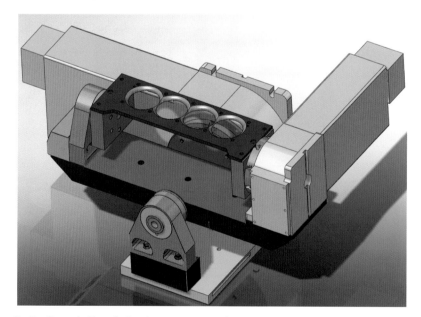

Figure 2-9 *Specialized dual rotary mechanism used in engine manufacture.*

Dedicated **Table/Table** machines are very capable of doing indexing/positioning work and are equally capable of simultaneous work. The inherent differences between the two are worth mentioning.

The **Indexing method** holds the workpiece much more rigidly than it is held for simultaneous machining work because the rotary axes are locked when machining. When rotating an axis, the rotary axis must first be unlocked with a designated M-Code. The axis is then rotated, and it is locked with another M-Code before machining is resumed. This sequence allows machining to be done in the machine's most rigid state.

When using **simultaneous** milling techniques, all the brakes must be disengaged, which will put the machine in its loose mode. For this reason it is always a good idea to use (when possible) indexing/positioning milling techniques for roughing cuts.

Machine Rotary Zero Position (MRZP)

Commonly, **MRZP** represents the intersection point of the two rotary axes, although sometimes the two rotaries may be offset by a specific distance. This distance must coincide or be relative to the part datum **PZP (Program Zero Point)** of the CAM system.

To accurately set up, operate, and program these machines, it is necessary to find the intersection of the rotary centers of the machine axes. Some, but not all, manufacturers have the values stamped on their rotary devices. However, those numbers are not to be trusted, and must be recalibrated regularly.

> **Finding the precise center of rotation is the foundation of accurate work.**

Even small discrepancies will magnify errors, further away from this machine rotary zero point.

Here are the steps to be taken:

1. Level the table by "zeroing" the indicator on either side of the table, as shown in Figures 2-10 and 2-11

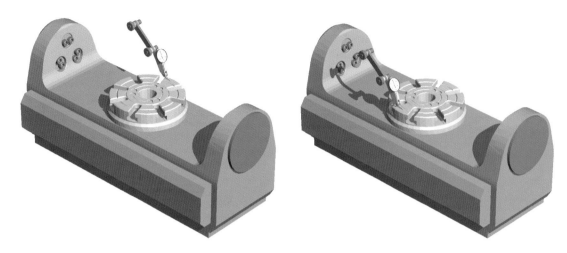

Figures 2-10 and 2-11 *Method of checking the level by dial-indicating both sides of the workholding table*

Figure 2-12 *Setting the dial indicator to zero before checking the level of the table.*

2. Find the XY zero, using the dial indicator. Zero XY and A at this point, as shown in Figure 2-13.

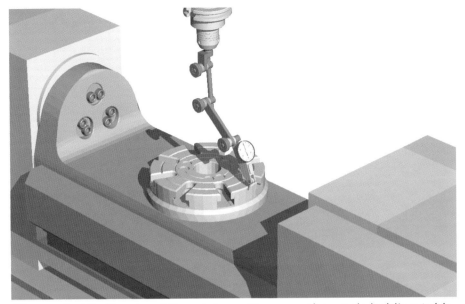

Figure 2-13 *Zeroing XY and A positions on the work-holding table.*

3. Rotate A+90 degrees and touch the OD of the table as shown in Figure 2-14.

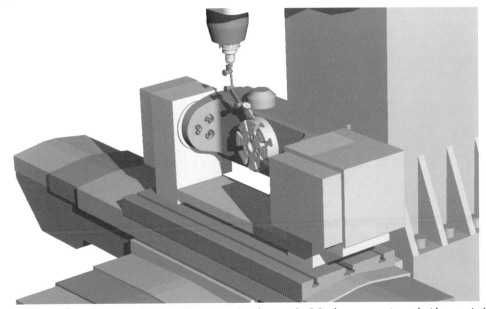

Figure 2-14 *After rotating the A axis through 90 degrees, touch the outside diameter of the table with the dial indicator.*

4. Rotate A-axis through 180 degrees from the previous position and make sure the indicator reads zero on the other side.

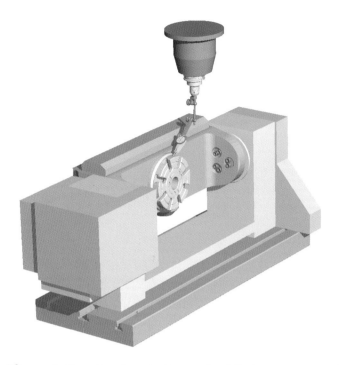

Figure 2-15 *After rotating the A axis through -90 degrees, touch the outside diameter of the table with the dial indicator.*

5. Move the Z-axis in minus direction the radius of the rotary table and set up a gage tower. The gage tower is used to set all the tool length offsets to Z=0.

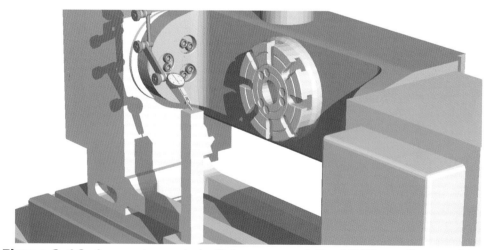

Figure 2-16 *A gage tower is built to represent the MRZP to allow tool length offsets to be set.*

This location is the machine's rotary zero position (MRZP), as illustrated in Figure 2-17.

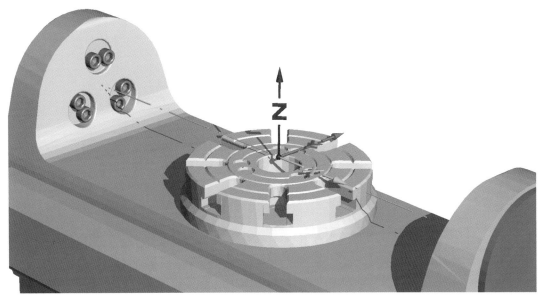

Figure 2-17 *The rotary zero position of the machine, as established by the outlined procedure.*

Note that the intersection of the dual rotary center lines is above the table in the example given. This location will be different for every machine, even from the same manufacturer. It is imperative that this position be checked regularly, especially after a heavy workload or a crash. Small misalignments can cause large errors because the tool position is measured from this intersection point.

All the **Active Coordinate Systems** also referred to as **Nesting Positions** or **Local Coordinate Systems**, for example G54 - 59, are relative to the **Machine Rotary Zero Point** (**MRZP**) position. It is good practice to set one of the nesting positions here, so that it will be captured in the Registry, allowing it to be recalled quickly, using **MDI (Manual Data Input)**.

For example: G90 G54 X0. Y0. A0. C0.

> The **PZP (Program Zero Point)** of the CAM systems must be set exactly to the **Machine Rotary Zero Point**, as seen in Figure 2-18.

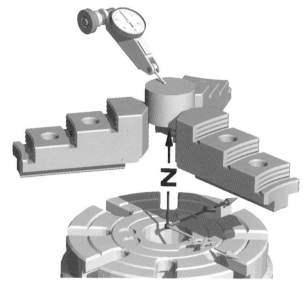

Figure 2-18 *Relationship between the MRZP and the PZP.*

Some CAM systems call this position the **World Zero, Master Zero**, or the **Origin**. The main thing to remember is to draw the part in the same specific position relative to this **World Zero** as it sits on the machine, relative to **Machine Rotary Zero Point**.

Nesting Positions

Nesting positions are widely used for positioning work. These positions, shown in Figure 2-19, are temporary **Active Coordinate Systems** and are typically set in relation to different faces of the part or fixture face, tooling ball, or dowel pin.

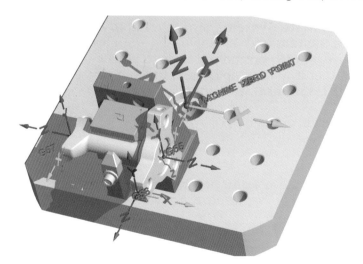

Figure 2-19 *Sketch showing some of the many local coordinate systems used in CNC programming.*

The advantage of using these **Local Coordinate Systems** is that you can easily follow the program on the controller's display screen because the absolute values shown there will reflect the values relative to each locally-nested position. Z+1.000, for example will be 1.000 (inch) above the part face.

Despite the fact that CAM systems all use different naming conventions for their coordinate systems, they all handle the local coordinate system in a similar way. Some of the names used by CAD/CAM systems include: **Part Datum, Active Coordinate System, Local Coordinate System, System View,** and **Tool Plane with an Origin**.

The disadvantage of using a number of different local coordinate systems is the potential for misalignment when picking up these positions manually with a dial indicator. Many programmers use only one coordinate system for 5-axis work. They use the **Machine Rotary Zero Point (MRZP)** as the part datum and let either the CAM system or the machine's controller calculate the special movements necessary. If a part is placed in the same position in the CAM and in the machine, the CAM is very capable of generating the correct code.

The advantage of using a single coordinate system is that the part needs to be indicated only once. The disadvantage is that it is harder to visually follow the program on the controller's display screen. The system will have to be switched over to **Distance to Go** for safer operation.

Using a real 5-axis machine as a verification system is inefficient, cumbersome, and very dangerous. There are many machine simulation software packages available that can save a lot of time and money, and these are covered in another chapter.

Rotary Table Dynamic Fixture Offset

The Problem
CAM generates code for a given position of the program zero point (**PZP**) relative to the center of rotation machine zero point, (**MRZP**). The machine operator may run the code later, on the night shift, at a different location **APZP (Actual Part Zero Point)**. He or she may not be able to place the part exactly where the CAD/CAM programmer intended it to be. If the operator does not have the access or the ability to make the change, then the job will have to wait for a reposted code to be supplied.

Modern CAD/CAM systems can easily calculate new code if the part is moved. But as previously mentioned, the part will have to be moved to exactly the same position in the CAM system and then the code will have to be recalculated.

The Solution
If the operator doesn't have access to the CAM system, and is unable to match the CAM's part position on the machine, an option on the machine will be needed to compensate for the discrepancy between the two positions. This option is called

Rotary Table Dynamic Fixture Offset (RTDFO).

When the **Rotary Table Dynamic Fixture Offset** function is activated on the controller, the **Program Zero Point (CAM datum)** is offset to correspond with the set fixture offset amount, as shown in Figure 2-20. This offset is the distance between the center of rotation (**MRZP**) and the **Part Zero Point (PZP)** and it must also take into account the angle of the rotary table. This function is convenient because multiple-face machining can be executed by setting one point as the reference when machining a complex workpiece.

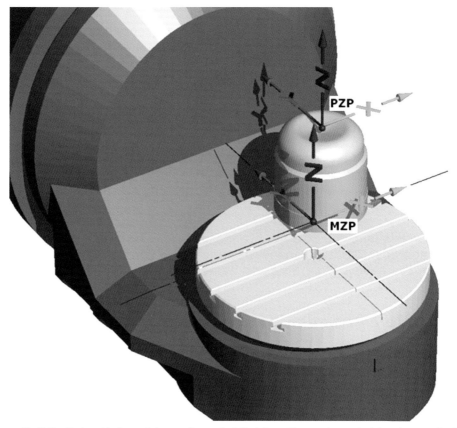

Figure 2-20 *Potential problems in establishing the rotary table dynamic fixture offset (RTDFO).*

There are 2 ways to use **RTDFO**:

1. Set the fixture offset amount manually on the Fixture Offset screen of the machine, illustrated in Figure 2-21.

▣ Function selection key 🖳 (OFFSET)
→ [FIXTURE OFFSET]

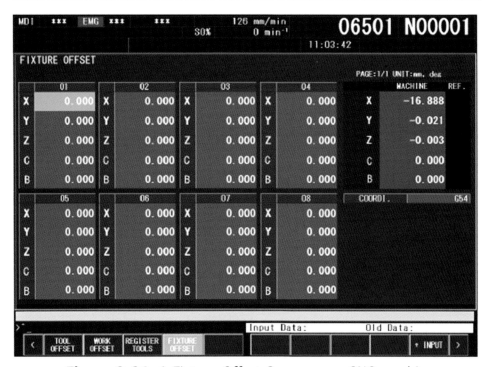

Figure 2-21 *A Fixture Offset Screen on a CNC machine.*

2. Specify the values in the machining program (G-Code).

The fixture offset amount is the distance between the rotational center (**MRZP**) and the workpiece zero point, used by the CAM program as the Program Zero (**PZP**).

> **G10 L21 Pn X_Y_Z_B_C_**
>
> **n**   **Fixture offset number (1-8)**
>
> **X_Y_Z_B_C_**   **Fixture offset amount for each axis**

When using the G90 mode, the specified values are set.

When using the G91 mode, the sums of the specified and the previous values are set.

Activating **RTDFO**:

> **G54.2 Pn;** **RTDFO – ON**
>
> **G54.2 P0;** **RTDFO – OFF**
>
> **n** **Fixture Offset number (1-8)**

The G-Code below shows an example:

```
%                                              %
O0001 ( PROGRAM - ZERO )                       O0001 ( PROGRAM - CLONE2 )
( DATE - 02-11-07 TIME - 07:22 )               ( DATE - 02-11-07 TIME - 07:22 )
G21                                            G21
G0 G17 G40 G80 G90 G94 G98                     G0 G17 G40 G80 G90 G94 G98
G91 G28 Z0.                                    G91 G28 Z0.
G28 X0. Y0. B0.                                G28 X0. Y0. B0.
G30 X0. Y0. Z0.                                G30 X0. Y0. Z0.
( TOOL - 31 DIA. OFF. - 31 LEN. - 31 DIA. - 16. )   ( TOOL - 31 DIA. OFF. - 31 LEN. - 31 DIA. - 16. )
( G43.4 G5 P10000 )                            ( G43.4 G5 P10000 )
T31                                            T31
M6                                             M6
G49                                            G49
G90 G0 Z500. B0. C0.                           G90 G0 Z500. B0. C0.
G43.4 H31 Z250.                                G43.4 H31 Z250.
G05 P10000                                     G05 P10000
G54                                            G54
/G10 L21 P1 X100. Y0. Z-100. B0. C0.           G10 L21 P1 X100. Y0. Z-100. B0. C0.
/G54.2 P1                                       G54.2 P1
M69                                            M69
M11                                            M11
G90 G0 X137.083 Y3.537 C-81.266 B18.081 S10000 M3   G90 G0 X137.083 Y3.537 C-81.266 B18.081 S10000 M3
Z188.477                                       Z188.477
Z93.477                                        Z93.477
G1 Z88.477 F2000.                              G1 Z88.477 F2000.
X135.846 Y3.258 Z88.776 C-81.64 B18.209 F6000. X135.846 Y3.258 Z88.776 C-81.64 B18.209 F6000.
X134.639 Y2.923 Z89.037 C-81.997 B18.323       X134.639 Y2.923 Z89.037 C-81.997 B18.323
X133.464 Y2.526 Z89.244 C-82.331 B18.413       X133.464 Y2.526 Z89.244 C-82.331 B18.413
X132.321 Y2.061 Z89.385 C-82.639 B18.476       X132.321 Y2.061 Z89.385 C-82.639 B18.476
X130.983 Y1.606 Z89.788 C-82.946 B18.652       X130.983 Y1.606 Z89.788 C-82.946 B18.652
X129.46 Y1.382 Z90.424 C-83.336 B18.934        X129.46 Y1.382 Z90.424 C-83.336 B18.934
X127.722 Y1.54 Z91.324 C-83.869 B19.336        X127.722 Y1.54 Z91.324 C-83.869 B19.336
X125.738 Y2.196 Z92.504 C-84.591 B19.874       X125.738 Y2.196 Z92.504 C-84.591 B19.874
X124.067 Y3.421 Z93.492 C-85.439 B20.334       X124.067 Y3.421 Z93.492 C-85.439 B20.334
```

Figure 2-22 *Example of G-Code data for setting RTDFO.*

If the machine does not have the option mentioned above, the CAD geometry will have to be moved, and the G-Code re-posted in the CAM system.

Note that the above example is for Fanuc controllers. Other controllers have a variety of names for this and similar functions.

Head/Table Multiaxis Milling Machines

As their name suggests, these machines have a rotary table and a tilting head.

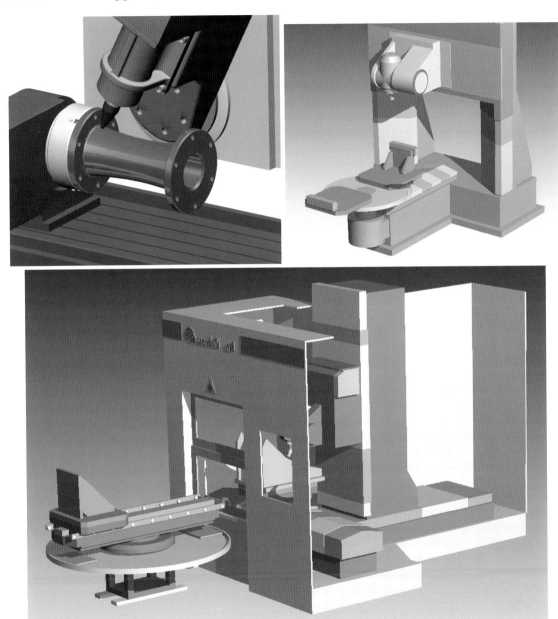

Figures 2-23, 2-24, and 2-25 *Example of Head/Table multiaxis milling machines, which have rotary tables and tilting spindle heads.*

2

Head/Table machines are arguably the most capable of the three groups illustrated and can machine large, heavy parts. On some machines, the rotary table can be supported by a steady rest and it rotates the part only around its own axis. The pivoting spindle head carries the weight of the tool. It needs to be capable of handling the cutting pressures as it is manipulating the tool.

These machines are also well suited for both indexing and simultaneous work. Some have the capability to calculate axis substitution internally, enabling the user to program parts in the 2D flat plane and then wrap the plane around a specified fourth-axis diameter.

How does axis substitution work?

Axis substitution is shown in Figure 2-26, and is effected by the following procedure.

- Measure the A-axis diameter and multiply it by Pi to find the circumference.
- Draw a rectangle where the Y side is the circumference and the X side is the length of the part.
- Create the cutting geometry inside this rectangle.
- Create a 3-axis toolpath, XYZ, and activate axis substitution by first defining the A-axis diameter.

On a Bostomatic controller, for example, this result is achieved by adding two lines of code.

G25 A3.000	**A-axis diameter**
G131	**Axis substitution Y to A active**

Figure 2-26 *A part produced by means of axis substitution.*

After these blocks are read, all Y-axis moves will be replaced by instructions for A-axis rotary motions. If the machine doesn't have this capability, this same process can be achieved with any modern CAD/CAM system.

The rotary axes on these machines usually have unlimited rotary motion. Some machines can even spin the workpiece as in a lathe. The secondary pivoting axis has an upper and lower limit. In order to accurately set up, operate, and program these machines, it is necessary to find the intersection of the rotary and the pivoting axes. Some examples of machines at the **Zero** position are shown in Figures 2-27, 2-28, and 2-29

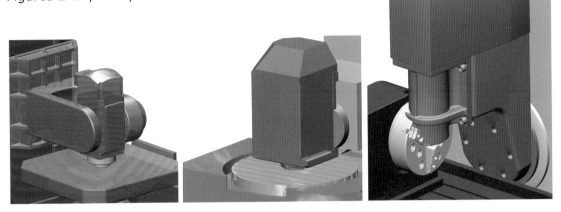

Figures 2-27, 2-28, and 2-29 *Examples of machines with spindles at the zero position.*

Note that without considering the tool, all these machines align the spindle face with the center of the rotary axis while the pivoting center point is some distance away from center. This distance is commonly called the **Pivot Distance**. The **Gage Length** is the distance from the spindle face to the tool tip.

The sum of the **Pivot Distance** and the **Gage Length** is the **Rotary Tool Control Point (RTCP)**, which has to be triangulated for every 5-axis position of the toolpath. Figures 2-30 and 2-31 show examples of B90 rotation with and without **RTCP**.

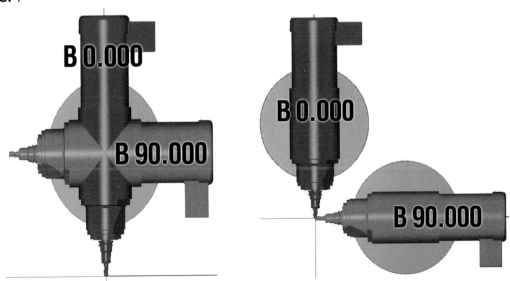

Figure 2-30 *Example of B90 rotation without RTCP, and* **Figure 2-31** *B90 with RTCP active.*

The machine's linear axes also have to move along the X and Z axes in order to keep the tool tip stationary in space as it executes the pivoting B90 motion. CAM systems will make the necessary calculations during "Post Processing." Some machines have the ability to calculate the necessary motions automatically, based on the offsets shown in Figure 2-31, captured in the machine controller's registries.

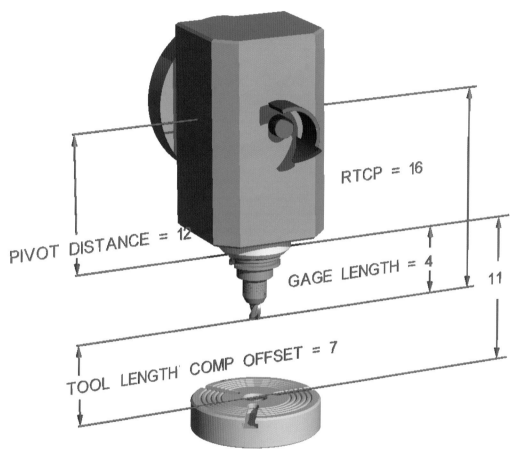

Figure 2-32 *Multiaxis offsets.*

Fanuc example:
G43.4, G43.5 5-AXIS ROTARY TOOL CENTER POINT CONTROL (RTCP)

If the **Rotary Tool Control Point** (**RTCP**) function is used in the Fanuc program, the spindle position is automatically adjusted in synchrony with all rotations, as shown in Figure 2-33 and the listed code lines beneath the figure. As a result, the relationship between the tool center point and the workpiece will always stay fixed.

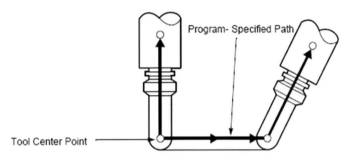

Figure 2-33 *The relationship between the tool center point and the workpiece stay constant.*

```
G90 G54 G00 X0 Y0 B0 C0;

S_M03;

G00 G43.4 Z_H_;

X_Y_B_C_;

Z_;

G49;
```

G43.4 . . .   Tool center point function (Type 1) ON

X,Y,Z, . . .   (G90) The coordinate value of the end point of the tool center movement

            (G91) The travel amount of the tool center

B,C . . . (G90) The coordinate value of the rotary axes end point

         (G91) The travel amount of the rotary axes

H . . .   Tool length offset number

G49 . . .   Tool center point control function (Type 1) OFF

Example:

G90 G00 G54 X0. Y0. B0. C0.; . . . Moves X, Y, B, C to **PZP**

S5000 M03

G43.4 Z1. H01; . . . Activate **RTCP**. Positions the tool tip at Z+1.000 while Z axis position is offset by offset data set for tool length offset number 1.

Some **Head/Table** machines will use both **RTCP** (**Rotary Tool Control Point**) and **RTDFO** (**Rotary Tool Dynamic Fixture Offset**) simultaneously. While RTCP is offsetting the tool position a combined distance from the head's rotary point (pivot distance + gage length), **RPCP** is compensating for the relative distance of the part from the **MRZP** (**Machine Rotary Zero Point**) to the actual fixture position.

If the machine doesn't have **RTCP**, to avoid repeated re-posting when tools are changed, it is common practice to pre-set all tools to the same length when possible.

Head/Head Multiaxis Milling Machines

All the rotary/pivoting motions are executed by the spindle head of the machine. These machines can be both vertical and horizontal, and they have limited motion. Some machines can change heads, not just tools. Heads can be straight, 90-degree, nutating, or continuously indexing. Some examples are shown in Figure 2-34.

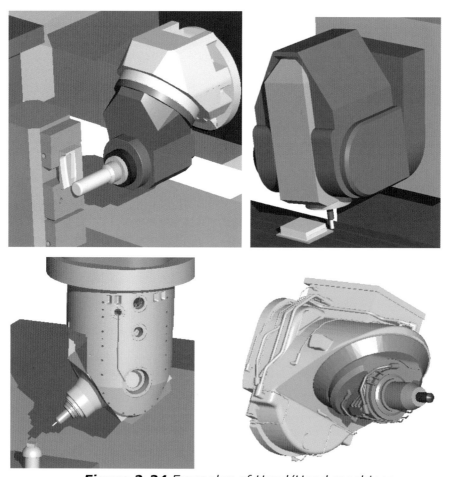

Figure 2-34 Examples of Head/Head machines.

All **Head/Head** machines have different behavior, based on individual installation settings. Rotary directions, limits, retractions, rotary wind-up, and handling singularities can all be altered from factory settings. The most important basic dimension needed is the rotary/pivot center point, which is measured from the spindle face to the head's rotary position. Machine manufacturers sometimes provide a nominal value, but it is essential that the manufacturer's value be double-checked, especially if it is a nice round number, for instance, 10 inches. The roundness is a good indication that the number is not accurate.

> **"Close" is not good enough. Knowing the exact dimension is vital if precision work is to be done.**

Finding the Pivot Distance

1. First, make sure that the machine head is in a perfect vertical orientation by touching the machine's table with a dial indicator, then rotating the indicator. The indicator should read zero around the whole circle as shown in Figure 2-35.

Figure 2-35 *Indicating vertical position.*

2. Place a 1.000 diameter dowel pin into the master tool holder with a known **Gage Length (GL)**.

3. Touch the dial indicator plunger as shown in Figure 2-36. A flat attachment helps here. Set the indicator to zero and record the Z value on the controller's screen. Let's call this value **Z maximum**.

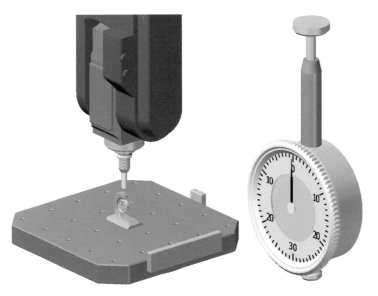

Figure 2-36 *Touching the dial indicator plunger is eased by having a wide, flat top on the plunger.*

- Do not move the machine on the X axis. Move only on the Y and Z axes. Move to a safe point on the Z axis, and rotate the A axis through 90 degrees into a horizontal orientation. Next, move on the Y axis in the plus and on the Z axis in the minus directions until you get to the position shown in Figure 2-37.

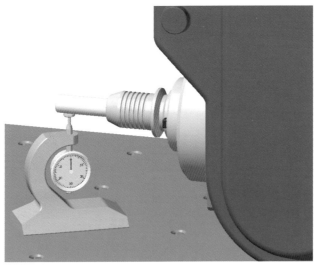

Figure 2-37 *Z minimum position.*

Record this Z value on your controller's screen and let's call this **Z minimum**.

You should have the following values handy:

> **Z maximum**
>
> **Z minimum**
>
> **GL - Gage Length**
>
> **R - Dowel pin radius = .5000**
>
> **Formula to calculate Pivot Distance:**
>
> **PD = Z max – Z min – GL + R**

This distance (**PD**), will be used by the post processor. Most CAM systems will drive the **Pivot Point** and they will have to calculate the tool tip location for every programmed position. The tool tip location is the **Pivot Distance** plus the **Gage Length** away from the **Pivot Point** at all times, and must be triangulated based on the rotary/pivoting angles. Even small discrepancies in the **Pivot Distance** will be magnified into large tool position errors in the final program.

4-Axis Machines

If a third-party, single rotary mechanism is placed on a 3-axis milling machine, it becomes a 4-axis machine. The most popular dedicated 4-axis machines are the horizontal types shown in Figure 2-38.

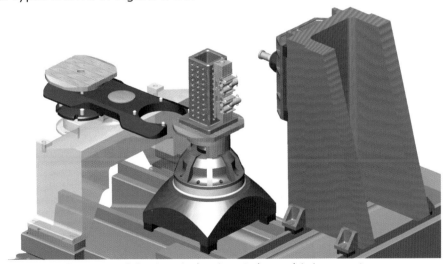

Figure 2-38 *4-axis horizontal machining center.*

These machines are mostly used for tombstone work, where parts are clamped to all sides of the tombstone fixture and machined by rotating them into different positions. The chips don't collect on the work-piece because they fall away by gravity and are cleaned off by strategically-placed coolant nozzles.

The example in Figure 2-38 shows a pallet changer, which is positioned outside the machine's enclosure, allowing the operator to load workpieces and unload finished parts during the machine cycle. Elaborate pallet changer assemblies are also available, with multiple tombstones on which a variety of different jobs can be pre-loaded and made ready-to-run. This arrangement allows for quick changeover to a new job without stopping the machine.

General Maintenance and Issues for Multiaxis Machines

It is recommended that all machine tools be kept clean and free of objects that can cause damage, and this rule is even more important on 5-axis equipment. Realigning routines should be done at regular time intervals, and most certainly after heavy work, overloads, or a crash. A log should be kept of the machine's vital statistics and operators should be instructed to listen for any new sounds coming from the spindle(s) or rotary mechanisms.

Some common problems include:

- Sometimes the rotary brakes will fail and they won't disengage. The rotary mechanism will then work extra hard to rotate from position to position, and if simultaneous work is in hand, the unit will eventually fail.

- Some dual rotary table center lines do not intersect. Some of these apparent discrepancies are by design, as shown in Figure 2-39, and some are not. If the apparent error is by design, it is usually a large number that can be seen by eye. If it is not by design, it won't be noticeable. It must not be assumed that the centerlines are in line, and they must be checked. Misalignment can be compensated for by inputting the relevant value into the post processor.

Figure 2-39 *Example of rotary mechanisms placed in offset positions by design.*

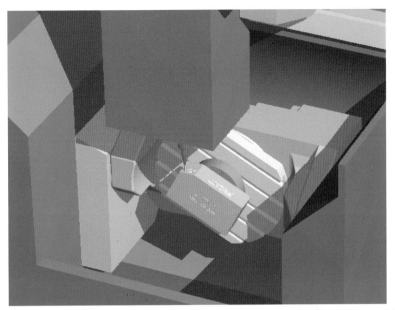

Figure 2-40 *These rotary mechanisms appear to be intersecting.*

- Some **Head/Head** types of machines will not run true. To check this aspect, arrange the machine with the secondary axis pointing down vertically as shown in Figure 2-41. Then, rotate the primary axis through 360 degrees. The dial indicator should read zero throughout this motion.

Figure 2-41 *Indicating run-out.*

The machine types described in this chapter are built by many different machine builders in a variety of sizes, shapes, qualities, and prices. The quality of a machine will be best highlighted when fast, simultaneous, multiaxis motion is being used. A good-quality machine will execute these motions quickly and repeatedly, in a smooth synchronized way, without one rotary axis waiting for another, and without backlash or vibration. The rotary mechanisms will have minimal run-out, and the rotary centerlines will align precisely. Cheaper machines may execute positioning movements well, but will execute simultaneous motions poorly.

Many manufacturers will list "positional repeatability" in their specifications because it is a good measure of the machine's quality. One way to do a quick check of repeatability is to set up an indicator on the machine table, engaging the tool holder at zero. Then a ten-minute routine involving all the machine axes with multiple rotational moves should be executed, terminating with returning to the start position. The indicator's readout relative to zero is the measurement of the positional repeatability.

It is not necessary to buy the most expensive machine, but only to take a good look at current and potential future needs when considering purchase of a multiaxis machine.

Milling Machines with Five or More Axes

Most machines with more than five axes are built for specific manufacturing applications. Some examples include those shown in Figure 2-42:

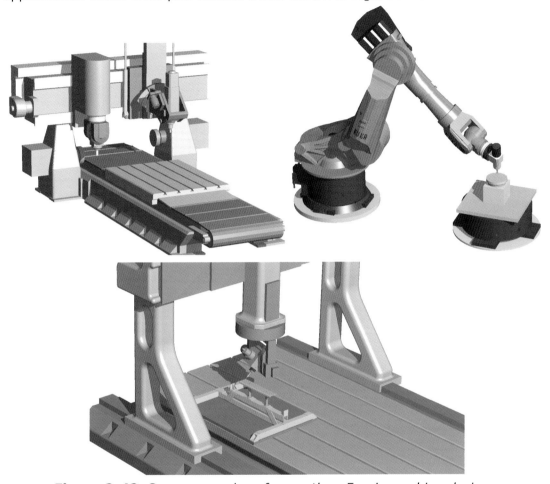

Figure 2-42 *Some examples of more than 5-axis machine designs.*

It is possible to assemble multiple 9-axis subsystems, and some manufacturers have built machines with over 100 axes. Many of these axes are part of elaborate work-holding systems, and have parts that need to be rotated out of the way of other machine components during some manufacturing processes. Many such machines are controlled with dedicated M-codes, which activate pre-set subroutines.

A simple example of such a subroutine is an M06, which causes a tool change. Observe closely what happens on any machine with an automatic tool-changer: the machine slides travel to pre-determined locations; the tool-change carousel advances the chosen tool; a little trap door may open, depending on the machine; then a swing-arm will exchange the tool between the spindle and the carousel. This whole choreography is just one of many internal macros, ready to be activated by a simple code like M06. On multiaxis machines, many more of these internal macros are available. Most of the time, the macros need to work in synchronization.

3

Cutting Strategies

If drawings of the same multiaxis part were given to five different CNC programmers, chances are good that they would come up with five different methods to machine the part. This variability is a product of experience, available multiaxis equipment, available CAD/CAM systems, tooling, fixturing, material, and quantities.

What does every CNC programmer do when asked to write a program for a new part? He or she will create a mental image of the part, and based on the above factors, go through a variety of different scenarios to determine how to machine it. These decisions will include how to hold the part, and which side to start on. The programmer will then mentally go through the whole process of removing all the excess material from the starting stock in order to free the desired part from within it. Most programmers will brainstorm repeatedly and come up with multiple solutions, eliminating the weakest ones, adding new ideas, and then making the final decision. This whole process happens long before the creation of the actual toolpath. This pre-work meditation is the single most important part of the whole manufacturing process.

The process described above is the same, whether 3-axis or multiaxis work is being considered. The big difference is usually with the fixturing. Work holding is among the first decisions to be made when programming a 3-axis machine. Many multiaxis programmers will place the part data on a virtual machine. This process lets them levitate the part in the air and simulate the machine's motions, without a fixture present, to see if all motions are possible without violating the machine's work envelope boundaries. The part will be moved in space to achieve optimized, synchronized motions. Final fixture placement, or design, might be one of the last steps.

Of course this procedure is not always possible, but when a fixture is predetermined, additional effort will be needed to make sure there are no collisions between the fixture, tool, shank, arbor, or tool holder. Avoiding collisions is a big part of multiaxis programming. Collisions can occur not only during cutting, but also during tool changes, pallet changes, or manual retraction moves after an abrupt program stop. For example, after a power failure, the tool could be in a position where the only safe retraction move is simultaneous multiaxis motions.

The single most important part of multiaxis programming is the initial time that is spent on deciding how to tackle the job. Machining sequences should be kept simple, not made complicated just because the shop has the latest equipment, the most powerful CAD/CAM system, or an unlimited budget. Here are some questions that need to be considered:

- **How many parts are needed?**
- **How much time is available?**
- **What is the material?**
- **What machine is available?**
- **How good is the CAD/CAM system?**
- **How well do you know CAD/CAM?**
- **What tooling is available?**
- **Do you have to use existing fixtures or can you make your own?**
- **Are there any special requirements?**

Limitations apply to every tool in the shop. The trick is to work around those limitations. The difference between a good multiaxis programmer and an average one is that the good one is industrious. If one approach doesn't work, another one will be tried until the best solution appears. Regardless of the CAD/CAM system in use, many times extra geometry will have to be created to achieve the best results.

Do the Prep Work
The time invested in preparing the work will be invaluable in the long run. Once a decision has been made on how the job will be handled, it is important to organize the work. Divide up the operations in the CAD/CAM system and move necessary geometry to easily-recognizable named layers/levels. This preparation will make it possible to isolate individual features and allow a focused workflow.

Make a Tool List
It is very important to make a tool list for any job. Start by analyzing the part geometry diligently. Find the smallest fillets. Measure how much room there is between features to determine the minimum and maximum tool diameters that can be used. Check what tools are readily available in the shop to see if any of them can be used, especially if you are already familiar with their performance. If you must order tools, do some research on their performance and availability.

Determine Fixturing

Check on available fixtures, vises, and clamps. Use existing vises and fixtures whenever possible, to keep the costs down. The equipment should be modeled in the CAD/CAM system and organized into libraries that can be readily accessed and loaded for virtual simulation when checks are made for possible collisions.

Compare Machines

If more than one machine is available for the job, some comparisons should be made. Among essential checks are: work envelope limitations, maximum RPM, feed-rates, and controller capabilities.

Know Your Stock Options

Material stocks must also be considered. If the material is unfamiliar, some research will be needed on different cutting characteristics. The original form may be a billet, a cylinder, a casting, or a forging, and may require some preparatory work before machining can start.

NOTES:

Indexing Multiaxis Toolpaths

Set-ups using indexing or indexed work are rigid and precise. Other common names used for such set-ups are 2+3 machining or positioning, and fixed rotary work. With indexing work, the rotary/pivoting axes are used only for positioning, and cutting (machining) takes place with only the three linear axes moving. Indexing work is the "bread and butter" of the multiaxis machining industry. Many parts are mass-produced by this method, and it is the most basic multiaxis concept. It is an easy transition from multiple set-up, 3-axis work to a single set-up indexing one. The graphics in Figure 4-1 show how one part can be cut from many different angles without being removed from the fixture.

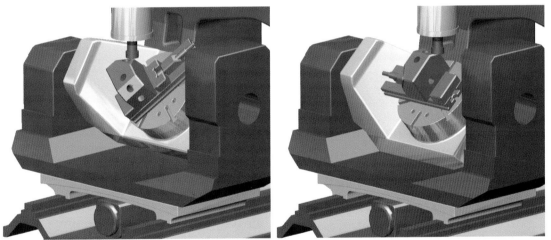

Figure 4-1 *Images showing how one part can be cut from many different angles, without being removed from the fixture.*

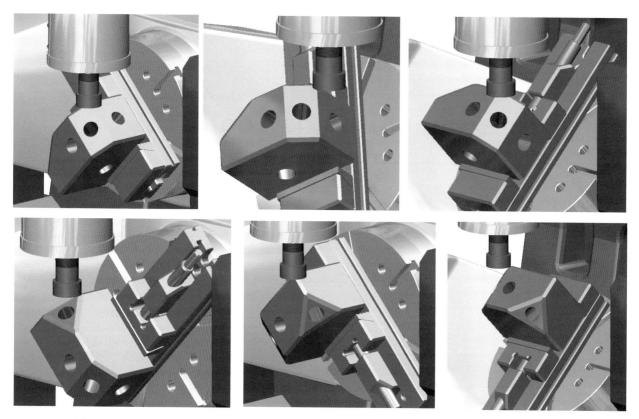

Figure 4-1a *Images showing how one part can be cut from many different angles, without being removed from the fixture.*

Figure 4-2 *Part of an aircraft landing gear machined with an indexing set-up.*

The concept may be simple, but it allows for the manufacture of very complex parts with precision, like the samples shown in Figures 4-2 and 4-3.

Figure 4-3 *An aerospace component machined with an indexing set up.*

Indexing Methods

There are many different indexing methods, and they can performed with equipment as simple as a manually-operated, custom indexing fixture. Third-party autonomous rotary devices also are available, which will execute pre-programmed indexing sequences at every cycle. The cycles can be activated manually or through a dedicated M-Code. If one of these methods is used, great care must be taken to synchronize the manual operations with the NC-code. Ample opportunities exist to make a mistake with these methods.

Figures 4-4 and 4-5 show two examples of custom indexing fixtures.

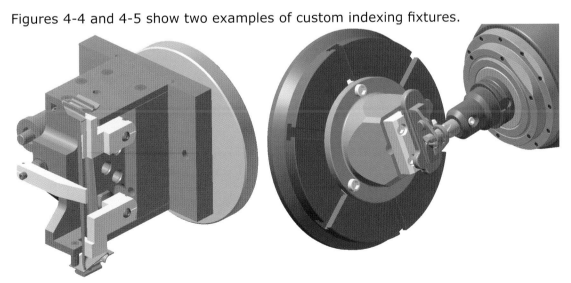

Figures 4-4 and 4-5 *Two examples of custom-built, indexing fixtures.*

The best method is to use fully-integrated, third-party, rotary devices, which will execute rotary commands directly from the NC-code. For these methods, the rotary pivot center must be precisely located (as described in Chapter 2).

Figures 4-6 and 4-7 show some examples of dedicated third-party rotary mechanisms.

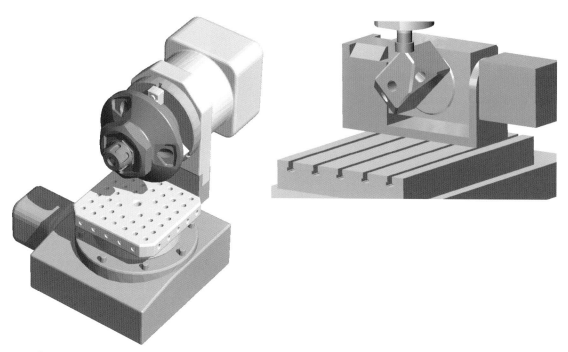

Figures 4-6 and 4-7 *Examples of dedicated third-party rotary mechanisms.*

The best approach is to use a dedicated multiaxis machine, if one is available. These machines have brakes on their rotary/pivoting axes, which provide extra rigidity during cutting. Typically, these brakes are released while positioning changes are made, but once in position, they are re-engaged so that the machine can stay in its most rigid state for cutting. Some machines are not numerically controlled but are capable of indexing only in certain increments (for example, 1 degree), and they often operate by lifting away from a serrated dividing plate during indexing.

Figures 4-8 and 4-9 show some examples of dedicated multiaxis machine, rotary mechanisms.

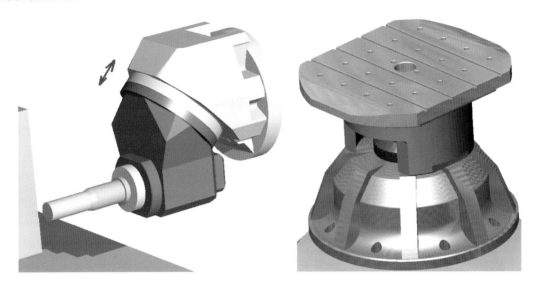

Figures 4-8 and 4-9 *Some examples of dedicated rotary machine components.*

On some machines, spindle heads can be changed repeatedly between operations. The examples shown in Figure 4-10 can be straight, set at a specific angle, or even adjusted steplessly to various angles.

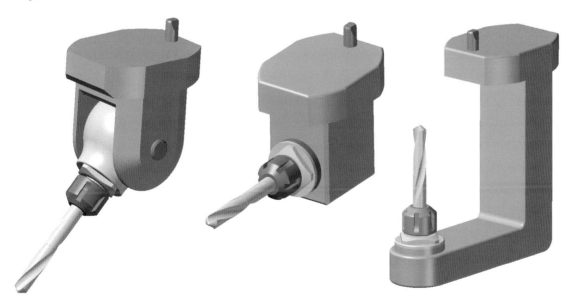

Figure 4-10 *Spindle heads on some machines are designed to be straight, set at a specific angle, or even adjusted steplessly to various angles.*

Other machines, used mainly in the medical and aerospace industries, are designed to index and hold the part with gripping axes while machining. Examples of these types of machines are shown in Figures 4-11 and 4-12.

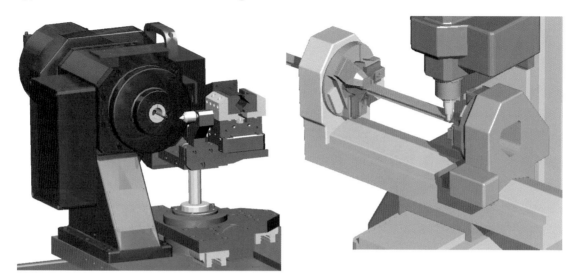

Figures 4-11 and 4-12 *Some machines are designed to index and hold the part during machining.*

Plain indexing is a very efficient way of moving parts into position for machining, especially when it is combined with pallet-changing. A pallet changer can be as simple as a single rotary indexing mechanism. It can also be as complex as a multi-pallet conveyer, with not just one, but multiple jobs, running in a pre-organized sequence. These systems are so flexible that a brand new job can be introduced into the queue without stopping the machine sequence, as shown by the examples in Figures 4-13 and 4-14.

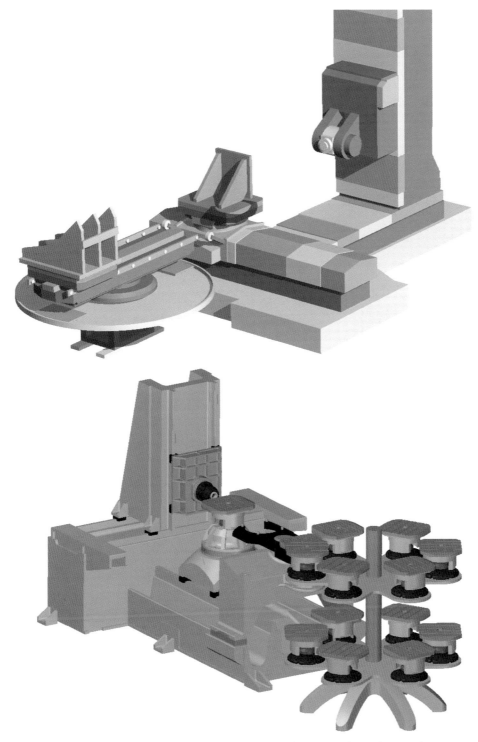

Figures 4-13 and 4-14 *Brand new jobs can be introduced into the queue without stopping the sequence with these pallet-changing machine designs.*

4

How CAD/CAM Systems Handle Indexing Work

Before discussing CAD/CAM system applications, it is important to establish some core understanding of how CNC machines work.

Prior to the invention of CAD/CAM systems, G-Code needed to be generated "by hand." Indexing work was handled just like any other programming job, the only difference being that another one or two axes were sometimes added to the mix. Most machine controllers have the ability to work in multiple local coordinate systems, also known as nesting positions. These local coordinate systems were, and still are, used in a variety of ways. One of the simplest ways is to place multiple fixtures and parts on the machine, establish the part data for each individual part, and assign individual local coordinate systems, as shown in Figure 4-15.

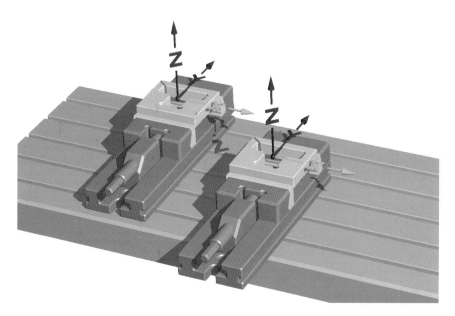

Figure 4-15 *Positioning two fixtures with parts on a machine and assigning individual local coordinate systems.*

The above example shows only two positions. The part programs would be the same for both, except that the local coordinate system designation would be provided at the beginning of the NC program (for example, G54 or G55 Fanuc). Different controllers use different designations for these nesting positions, but they all work on the same principle. Depending on the controller, numerous nesting positions can be designated.

This nesting concept is one that many people struggle with, and its understanding is key to multiaxis machining and programming practices. There are a variety of controllers and machines available that use this same concept, but they use different terminology to describe it.

Machine Coordinate Systems

Machine Home Position

Simply put, **Machine Home Position** is the center of the machine's universe. Every axis will travel to its **Home** (end of travel) position and the machine will stop there. At this **Home** position, in the machine's **Absolute Coordinate System**, all axes are reading/displaying **zero**. Every move the machine axes make from here will be relative to this **zero**. Every position captured, such as a nesting position, will be a relative position in the **Machine Coordinate System**. Every time a tool is changed, the machine will go to the pre-determined position specified in this **Machine Coordinate System**.

To establish the nesting positions, the first vise is loaded and checked to make sure it is square and secure before it is clamped in position. Then the workpiece is placed in the vise and the vise is tightened to hold the workpiece in place. Using an edge finder, the center top of the part is located, as shown in Figure 4-15. The machine's absolute position display should now show how far the axes are from the **Home** position. This position must now be captured and the machine must be made to remember this location. Machines remember by storing the relative distances (from **Home**) in their registry. How nesting positions are 'captured' depends on the type of machine and controller in use.

Active Coordinate System

Nesting positions on a multiaxis machine can be moved and rotated in one plane. They can also be rotated about the machine's rotary/pivoting axes.

> **The Machine Home Position is the center of the machine's universe.**

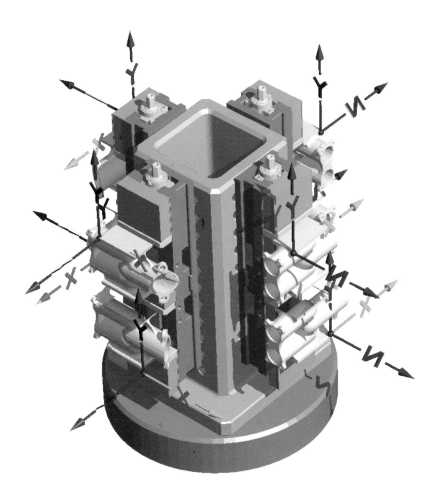

Figure 4-16 *Multiple nesting positions on a tombstone fixture.*

There are two popular ways to use nesting positions, the first of which is shown in Figure 4-16, which illustrates a tombstone fixture in use. Every part datum on the tombstone fixture shown has its own local coordinate system assignment. Many programmers feel that the arrangement shown in Figure 4-16 is the best way to use nesting positions.

The other way is to assign just one central coordinate system to the whole job as shown in Figure 4-17.

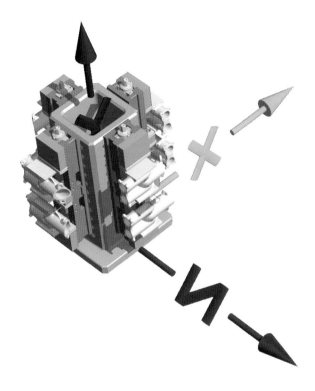

Figure 4-17 *Central coordinate system on a tombstone fixture.*

Both methods are correct and it is simply a matter of personal preference as to which one is used.

When it comes to machining a single workpiece, a preferred method is to use only one **Active Coordinate System** method, but this also is just a matter of preference.

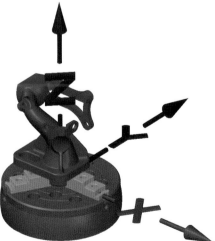

Figure 4-18 *Central coordinate system on a single part.*

Using a single **Active Coordinate System** requires that only one position be indicated on the machine. This approach simplifies the process and lessens the possibility of error.

Machine Rotary Center Point

So far it has been established that every machine has its own **Home Position**, which is its center of the universe. Every local coordinate system is a relative location in that universe. Also, the intersection of the rotary axis, commonly known as the **Machine Rotary Center Point**, is a relative location in that same universe, and its position is stored in the registry.

CAD/CAM System Origin

Every CAD/CAM system also has its own universe. They all have a **World Zero**, **Master Coordinate System**, **System Origin**, and so on. Just like machine tools, all these locations are called by different names. One thing you can be sure of — none of them will have the same **Home Position** as any other machine. The job of a CAD/CAM user and CNC machine programmer is to align the worlds of both the machines and the CAD/CAM systems.

If the **One Zero** method — where the local coordinate system on the machine, which is the **Machine Rotary Zero Point** — is in use, it is possible to simply match the CAD/CAM System's **World Zero** with that location. The part must then sit in the same relative location and orientation from the **Machine Rotary Zero Point** of the system and the machine, as seen in Figure 4-19.

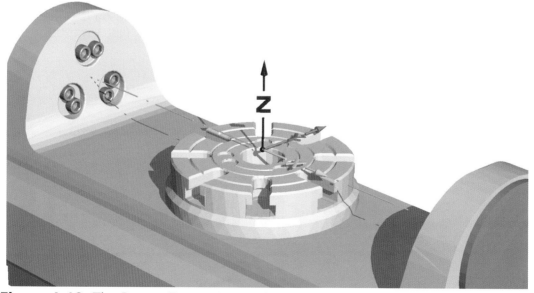

Figure 4-19 *The Rotary Zero Point is where the two rotary center lines intersect.*

If, on the other hand, the multiple nesting position method is preferred, new **Active Coordinate Systems** must be created in the CAD/CAM system as shown in Figure 4-20.

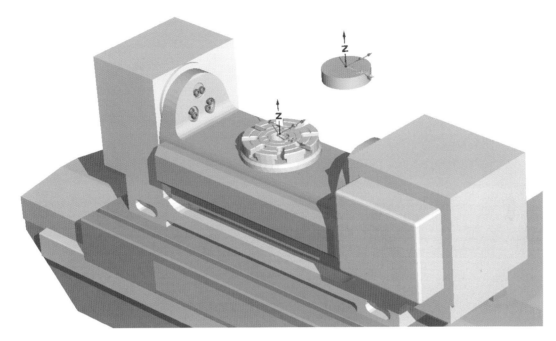

Figure 4-20 *The relationship of the part zero to the Machine Rotary Zero Point.*

Synchronizing Machine and CAD/CAM Coordinate Systems

These **Active Coordinate Systems** are the equivalent of the nesting positions (for example G54-59) on the machine. Different CAD/CAM systems establish active coordinates in different ways, as shown in Fig. 4-21. For the sake of simplicity, the following description will be kept very general.

An **Active Coordinate System** can be established by choosing an entity, such as a solid face, an arc, two lines, normal to a surface, normal along a line, or normal to a plane.

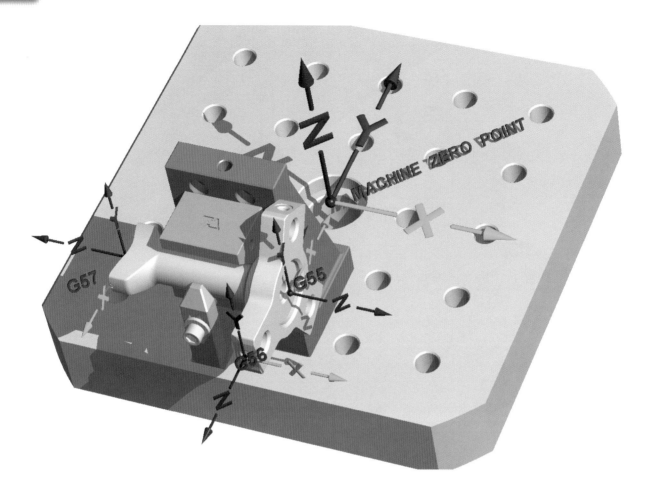

Figure 4-21 *Multiple local coordinate systems.*

One of the differences between programming a 3-axis machine and a multiaxis machine is the determination of where the fixture and part will be located on the machine table.

On a multiaxis machine, exact instructions must be given as to where the part should sit relative to the **Machine Rotary Zero Point**. As always, a bit of pre-planning will go a long way. Avoiding collisions between tools, tool-holders, fixtures, and machine components, for example, will be one of the major preoccupations. Creating an accurate library of the fixture plates, vises, clamps, tools, and tool-holders in use in the plant will help greatly in avoiding those potential collisions. Find the **Machine Rotary Zero Point** (*described in Chapter 2*) for every machine in the shop, and place the fixtures on those virtual machines in the CAD/CAM system. It is not necessary to model the whole machine, but at least the machine's table should be modeled. Extra care should be taken that all the models sit in this aligned universe (CAD/CAM and machine).

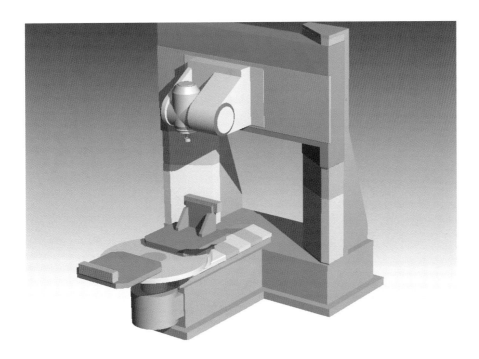

Figure 4-22 *Complete Machine Simulation.*

Depending on the CAD/CAM software selected, it is also possible to model and simulate the whole machine like those shown in Figures 4-22 and 4-23.

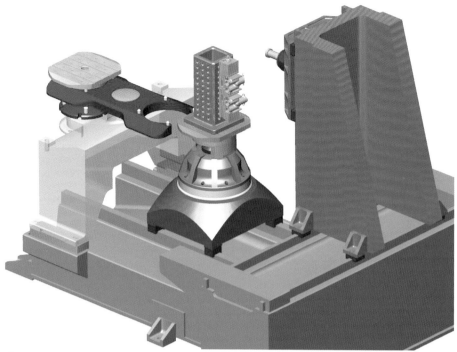

Figure 4-23 *Virtual 4-axis horizontal machine for simulation purposes.*

It is vitally important that the "business end" of the machine be modeled accurately, if any simulation is to be useful. By the business end is meant the head, fixture, table — in other words, the parts that can actually collide. Simulation will be discussed in more detail in a later chapter.

5

Simultaneous Multiaxis Toolpaths

Many people think that simultaneous multiaxis is the true form of 5-axis machining, when in fact, it is not necessary for all the machine axes to move at the same time for the machine to be considered 5-axis. Even a simultaneous 2-axis, rotary cutting motion may be considered to be a multiaxis toolpath.

> **Simultaneous** multiaxis machining is also known as **Continuous** 5-axis or **True** 5-axis machining.

The illustration in Figure 5-1 shows a 2-axis machine cutting a pattern onto a bowling ball. This machine only has a tilting B and a rotating C-axis. There is no Z axis. Instead, that motion is controlled by a software M code, which has an ON and OFF state — either lowering the tool onto the part, or lifting it to its reference position.

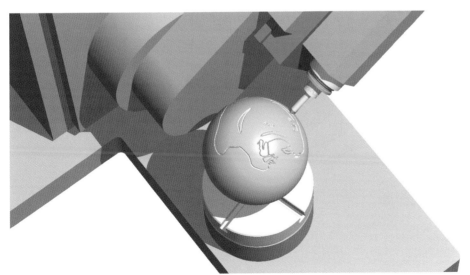

Figure 5-1 *Set-up on a 2-axis machine for engraving a bowling ball.*

The example in Figure 5-2 also shows a simple multiaxis motion — so simple that it can be programmed by hand. The program contains the following codes:

```
G01 Z2.0000 F90.

X-5.5 A2880.000 F50.

G00 Z5.
```

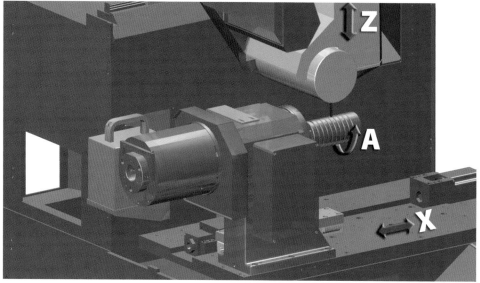

Figure 5-2 *A simple multiaxis set-up.*

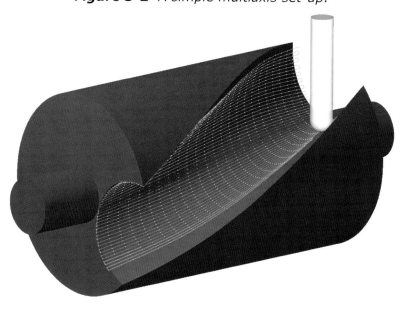

Figure 5-3 *Sketch of simultaneous cutting on a 4-axis machine -XYZA.*

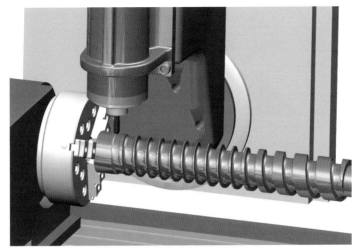

Figure 5-4 *A 4-axis machine set-up for cutting a variable-pitch thread on an auger using motions on XYZ and A axes.*

Simultaneous cutting on a 4-axis machine is shown in Figure 5-3, and a set-up for cutting a variable-pitch thread on an auger using 4-axis motions XYZ and A is shown in Figure 5-4.

Figure 5-5 illustrates a set-up on a similar machine, combining simultaneous motions, and using a flywheel to produce a knee-joint component using the 4-axis motions XYZ and C.

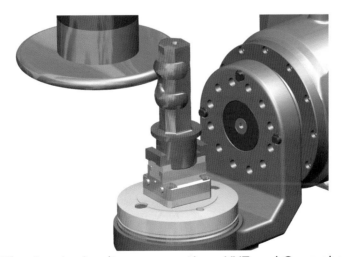

Figure 5-5 *The 4-axis simultaneous motions XYZ and C are shown cutting a knee-joint, using a fly-cutter.*

Many parts would be impossible to machine without simultaneous multiaxis motion. In the early days of multiaxis machining, many parts were designed around motion instead of as freeform CAD models.

An example is the spiral bevel gear shown in Figure 5-6, which would normally be produced on a special gear-cutting machine in an automobile plant.

Figure 5-6   *Spiral bevel gear produced on a 5-axis CNC machine.*

This gear was machined with the following manually-generated, motion-driven codes:

```
O0001
G20
G90G00X-3.75Y0.Z25.B-35.C0.
T1M06
S300M3
S3000M03
G43Z3.5 H1
Z3.25
G1Z2.9F200.
M98 P3000 L30
G90G00Z25.M05
M30

O2000
G91G1Z-.1F50.
X2.2Z-.1C60.B5.  (4-axis simultaneous motion)
X-2.2Z.1C-60.B-5.
M99

O3000
M98 P2000 L3
G91G00Z.3
Z1.
C12.
Z-1.
M99
```

This last example is very simplistic, but with some creative use of branching/looping logic. Some shops have used this technique to produce very complex parts.

There has always been a separation between design and manufacturing. Typically, part designers are not CNC programmers or operators. As a result, many designs don't take account of clean tool motion, or they include features that are hard to machine and require additional operations. In well-run shops, designers and production engineers work in conjunction, from the design process through to manufacturing. This is an ideal solution, but unfortunately not the norm. Working in conjunction, engineers can save many hours of valuable manufacturing time, tooling, fixture design and building.

CAD systems have evolved drastically and, as a result, it is possible to design and manufacture ever-more complex parts like the examples shown in Figure 5-7.

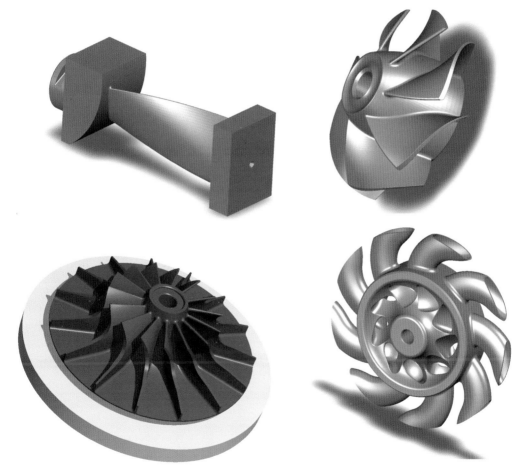

Figure 5-7 *Examples of parts produced on multiaxis milling machines, including turbine blades and rotors, impellers, pump components, brackets, and manifold covers.*

Figure 5-7 *Examples of parts produced on multiaxis milling machines, including turbine blades and rotors, impellers, pump components, brackets, and manifold covers.*

Developing cutting strategies for these multiaxis parts entails more than just creating toolpaths. The strategy is all about control. The goal is to create a toolpath that causes the smoothest, most efficient, machine motion inside the machine's "sweet spot" (the optimum work envelope), while avoiding near-misses and collisions between machine tool components, fixtures and holders.

The Optimum Work Envelope

The optimum work envelope is the space in which the machine's rotary axes rotate about the same diameters. The following is an example.

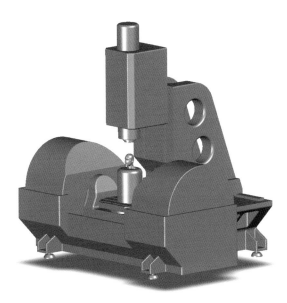

Figure 5-8 *A vertical milling machine with a trunnion-type dual rotary table, set up to machine a model of a human head.*

Machining of a model of a human head on a trunnion-type dual rotary table is shown in Figure 5-8. The head is high above the **Machine Rotary Zero Point**, measured along the Z-axis, but it is very close to the C-axis center point of the rotary table, measured along the X and Y axes.

In programming such a job, it is best to avoid creating simultaneous rotary cutting motions involving the full range of the tilting B-rotary axis (-15 and +105 degrees) while the C-axis is being rotated around its axis. Doing so will create uneven motions between the rotary mechanisms.

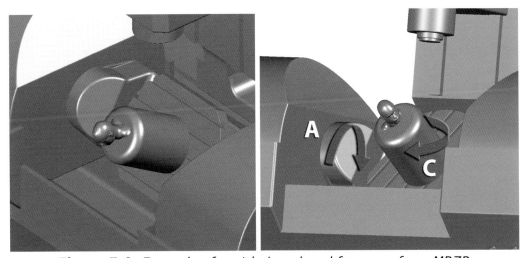

Figure 5-9 *Example of part being placed far away from MRZP.*

In Figure 5-9 it should be noted that the B-axis move is much longer than the C-axis move, even though the angular values are the same. The reason, of course, is that the circumferences are widely different for the B and C motions. High-quality machines handle these kinds of uneven rotary motions better than lower-quality machines because they synchronize the two rotaries to arrive at the same point, while maintaining a constant feedrate. CAD/CAM systems can also control feedrates by using **Inverse Time Feedrate** output. A more detailed overview of these controls is included in the Feedrates section of this chapter. At this point, it is sufficient to know that it will be much better to place the workpiece closer to the same rotary diameters of the specific machine, as shown in Figure 5-10, especially if a third-party dual-rotary table, or a lesser quality multiaxis machine are in use.

Figure 5-10 *The part is placed close to the Rotary Zero Point of the machine.*

Placing the workpiece close to the same rotary diameters of any particular machine, as shown in Figure 5-10, might not always be possible. But when it is, take advantage of this simple technique to better control motion.

Feedrates

On a 3-axis (non-rotary) machine, there is no need to specify a feedrate mode because these machines all operate in the units/time mode.

For example, if you designate a position as G91 G1 X7.07107 Y7.07107 F10, your machine slides will move the workpiece in a coordinated linear motion from its current position to an incremental destination of X7.07107 Y7.07107 at 10 inches a minute. The machine will move the workpiece exactly 10.000 inches in a straight diagonal line.

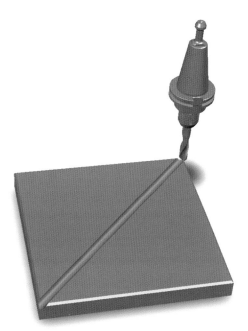

Figure 5-11 *A diagonal groove is machined by moving both table slides simultaneously using linear interpolation.*

With linear interpolation, the workpiece won't get to 10 inches/per minute instantly because the slides need to accelerate from zero. Once a speed of 10 inches a minute is reached (if the machine is capable), it won't instantly stop at its destination. Instead, the slides will decelerate to that position, but for this example, those lost times are negligible. We can calculate the time of this 10.0000 inch move with this equation: **10 inches/minute = 1 minute**.

Figure 5-12 *Circular interpolation is used to move the workpiece in a circular path.*

A planar circle cut using a `G3 I-5. F10.` command is illustrated in Figure 5-12. The resulting motion appears to be a true circle, but it is not. Any machine that has the standard three XYZ linear axes cannot cut a true circle; only an approximate one. The slides on these machines can move only in straight lines. Therefore, in order to generate a circular path, the controller will have to interpolate a circular move by breaking the programmed circle into a number of straight-line segments. On most machines, the circular tolerance can be set from inside the control parameter settings. The larger the size of straight-line segments, the less accurate the circles will be. A smaller number will result in more accurate circular cuts.

Changing the circular tolerance affects not only the circular accuracy, but also the feedrate used for the cut. The machine will have to slow down in order to maintain the accuracy set, and the feedrate will change based on the size of the arc. Large arcs can be cut with a faster feedrate than small ones.

Every quadrant of an arc includes a peak error area, which consists of the points where the linear axes intersect the arc at 0, 90, 180, and 270 degrees. As the machine interpolates the circle, it needs to reverse the slide motion of its linear axis to travel in the opposite direction. Even if a high feedrate is programmed, the machine's controller will limit the executed feedrate based on the circular tolerance set in the controller and the arc size currently being executed. For this reason, calculating cycle times is not an exact science.

> Multiaxis machines work with two types of feedrates:
> - **Standard** (G94 units/time), as described above
> - **Inverse Time** feedrate (G93)

Inverse Time Feedrate

During simultaneous multiaxis rotary motions, both rotary and pivoting axes must ideally arrive at a specified rotary destination at the same time. Otherwise, movement on one axis will stop to wait for the other rotary axis to arrive. This wait will cause the tool to dwell in one position, which in turn, will change the cutting pressure and deflection. In the best case scenario, this delay will cause an unwanted tool mark on the part surface. In the worst case scenario, the pause can even gouge the part. CAM systems handle this problem by linearization, which breaks up these moves into smaller segments and applies controlled **Inverse Time** feedrates to them.

The feed/minute is specified when the tool needs to move at a specified feedrate to maintain the necessary feed per tooth to cut the material consistently. To move the tool with that feedrate, the rotary center points need to move much faster in space, especially if longer tools versus shorter ones are being used.

The example shown in Figures 5-13 and 5-14 has only one rotary motion combined with X and Z linear moves.

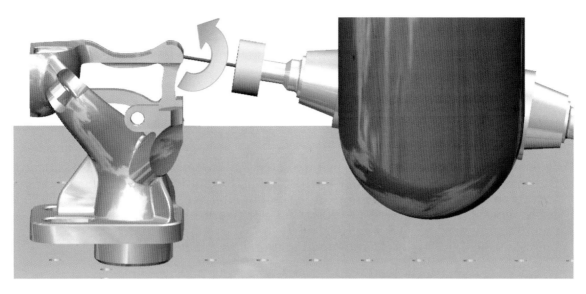

Figure 5-13 *The start position for machining a complex part.*

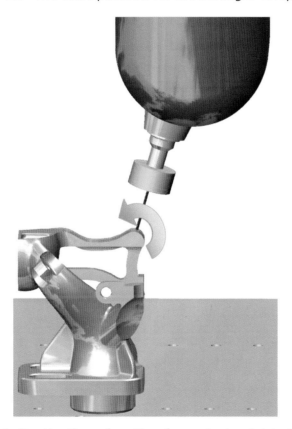

Figure 5-14 *Destination of motion from start point in Figure 5-13.*

Looking at the two illustrations in Figures 5-13 and 5-14, it is possible to observe and imagine the difference in travel distances between the tool tip and the rotary center point of the head. To maintain the programmed feed/minute on the tool tip, the center of the rotary spindle head needs to move very quickly. This scenario can be compared to runners on a track. Running in the inside lane of the track covers less distance than running on the outside lane of the track. The tool tip is the runner on the inside lane, and the center of the rotary is running on the outside lane.

In short, the machine should not be instructed to move from the current position to the destination at X units per minute. Instead, it should be told to move from start to destination, in X amount of time, in a smooth interpolated motion, on all the axes involved. On Fanuc type controls, G93 signifies the start of the inverse time mode. There must be an F command at the end of every line containing a G1, G2 and G3 code. The **Inverse Time** mode will not affect rapid G0 moves.

In **Inverse Time Feedrate** mode, an F signifies that the move between the current position and the destination should be completed in (1 divided by the F number) minutes. For example, if the F number is 2.0, the move will be completed in half a minute.

Inverse Time Feedrates were widely used in the early days of NC, but today many modern CNC controllers are capable of parsing standard feedrates into inverse time and vice-versa. (A parser is a compiler or interpreter). Usually, an inverse time smoothing algorithm is incorporated into this feature and it can be enabled, or disabled, in the controller's parameter setting.

Post Processors

CAD/CAM systems generate 5-axis vector lines along 3D paths. The 3D paths represent the tool motion as it follows the pattern being cut. The vectors represent the individual tool axis directions (IJK vectors) as the tool follows the 3D (XYZ) pattern. Every vector is represented by a line of code, and during toolpath creation, a resolution of these vectors can be specified, either by defining the minimum angular differences, or the linear distances between vectors. This information is written in a generic language. Depending on the CAD/CAM system, the language may be called APT, CLS, NCI, and others. Machine tool controllers do not speak or understand these generic languages, however they do understand many different languages and dialects.

The generic CAD/CAM code must be translated into a machine-readable language, a process that is called post processing. A post processor will calculate the axis motions needed on a specific machine to reproduce the CAM vector sequence. The post processor includes detailed information about the specific machine's physical and computing properties that allows it to generate the required accurate G-Code. This code, in turn, will govern the axis movements of the machine that are needed to machine the part. A different post processor will be needed for every type of multiaxis machine in the shop.

Post processors have built-in intelligence designed to detect rotary limits and automatically retract and reposition machine axes. Rotary moves are treated with a bias (not applying a neutral point of view correction to the process), based on the layout, as well as the primary and secondary rotary axes of the machine. Post processors will stay away, or warn of 5-axis instabilities, and they can output rotary rapid motions as programmed high feedrates to better control every aspect of a machine's motions.

There are always two possible solutions when a post processor maps a 5-axis tool orientation to a 5-axis machine tool's kinematics. The post processors will choose the best solution of the two. Consider the example shown in Figure 5-15.

The current position is XYZ A+80.000 B0.000. In theory, the tool could also reach this same position at XYZ A-80.000 B180.000, but that would be impractical because the part would be hidden from view and the operator would see the back side of the rotary device. Also, there is not enough Y-axis travel capability on this specific machine.

Figure 5-15 *One of the two possible solutions for a 5-axis position.*

Selecting the best 5-axis position is the task of the post processor writer. Another task of a post writer is to solve 5-axis instabilities, also known as pole singularities. These faults occur when the tool is vertical or almost vertical. Most posts will generate retract moves along the tool axis in these situations. Good posts will avoid erratic retract and large repositioning moves by tracking the possible angle pairs, angle change limits, and machine mechanical travel limits.

Many CAM systems handle safe motions between two subsequent toolpath operations with post processors. These controls retract the tool into a safe area, and a 5-axis machine repositions from one operation to the next. Instead of simply retracting to the Machine Home Position, safety volumes (box, hemisphere, cylinder) can be used for efficient tool retraction. Keep in mind that an efficient toolpath doesn't make erratic and unnecessary motions – it retracts the workpiece only to a minimum safe distance, and keeps the cutter engaged, while maintaining all machine axes in optimum positions.

Every CAD/CAM developer has dedicated departments devoted to writing and supporting their post processors, and there are many consultants making a living doing the same work. There is a great need for post processors because no two machines or operators are the same. Post processors can be customized, not only to suit individual machines, but also to suit the individual user's preferences. If a company wishes to attempt to modify its own post processor, most developers will provide training and documentation.

Developing a post processor for multiaxis machines takes a lot of effort, talent, professionalism, and perseverance. There are many "hackers" who are managing to "make it work," but a high-quality post processor is supplied with detailed documentation and user-defined switches.

An exceptional post processor writer visits corporate machine builders to get information directly, and then develops and tests the post processor on all the machine types in use. The post processor is thus tried, tested, and certified by both the CAD/CAM company and the machine tool builder.

6

Common Simultaneous Multiaxis Toolpath Controls

A good CAD/CAM system is one of the most important tools in a modern machine shop, and will provide enough control to confidently drive any multiaxis CNC equipment. The three major things that need to be controlled are:

- **Cut Pattern** - This pattern guides the tool's cutting directions.

- **Tool Axis Control** - The orientation of the tool's center axis in 3D space as it follows the cut pattern.

- **Tool Tip Control** - The geometry that the tool tip is compensated to follow.

In addition to those three major controls, which are defined in more detail in this chapter, good-quality CAD/CAM systems also provide additional collision-avoidance. This insurance will recognize the tool's cutter, shank, and holder. Different avoidance behaviors can be invoked when any of these components comes into proximity with the work-piece or a fixture. Different near-miss tolerances can be assigned to each of these tool components.

Cut Patterns

Cut Patterns guide the tool along specified paths. These patterns can be simple 2D or 3D wireframe, or solid primitives (for example, box, cylinder, and sphere.) **Cut Patterns** can also be complex multi-surface grids.

Some examples of cut patterns are shown in Figures 6-1 through 6-17

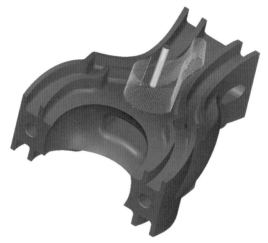

Figure 6-1 *Tool motion following a 3D curve projected on to the face of a workpiece.*

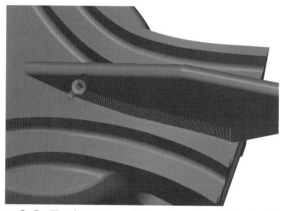

Figure 6-2 *Tool motion following the rib's bottom edge.*

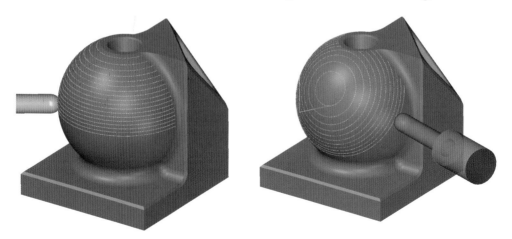

Figure 6-3 *A Cut Pattern is selected to slice the part in any given plane, for example, patterns 3 or 4.*

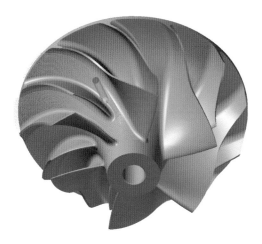

Figure 6-4 *Impeller floor surfaces that use a Cut Pattern that morphs between the two blade surfaces.*

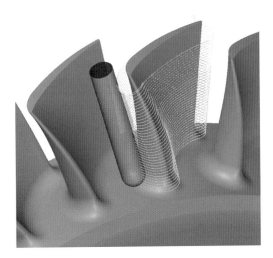

Figure 6-5 *Cut Pattern that is parallel to the bottom hub surface, while cutting individual blades.*

Figure 6-6 *The Cut Pattern for producing a cylindrical-spiral tool motion.*

Figure 6-7 *Cut Pattern produced by morphing between the two edge curves of the floor surface.*

Figure 6-8 *This Cut Pattern is shown by the red 3D curve projected on to multiple surfaces.*

Figure 6-9 *Floor surface being cut by morphing between two 3D curves formed by the floor's outer edge curves.*

Figure 6-10 *Cut Pattern parallel to the floor surface as it spirals down each blade.*

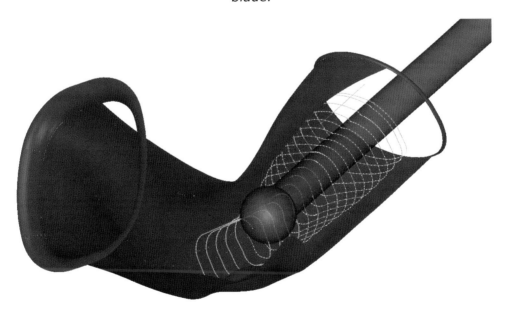

Figure 6-11 *Racing engine intake and exhaust ports machined with a spiraling Cut Pattern.*

Figure 6-12 *Path following a spherical Cut Pattern.*

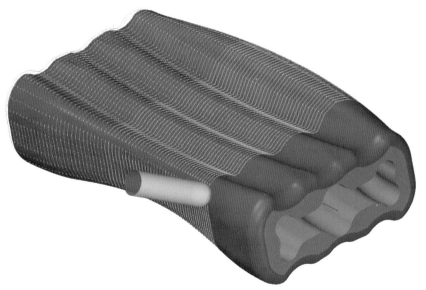

Figure 6-13 *Path following a box-shaped Cut Pattern.*

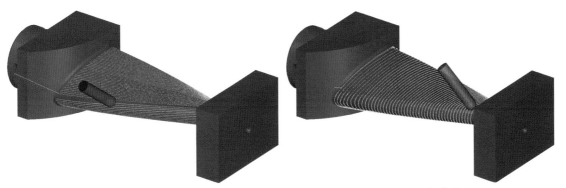

Figure 6-14 *Axial Cut Pattern on a turbine blade.*

Figure 6-15 *Radial Cut Pattern on a turbine blade.*

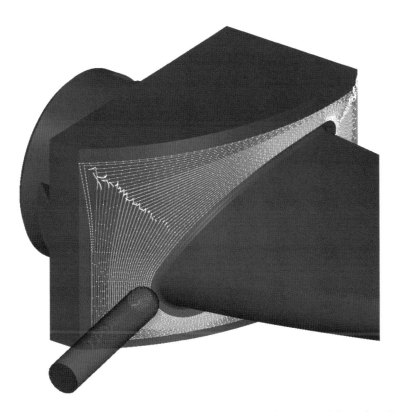

Figure 6-16 *Turbine blade's foot surface cut by morphing the Cut Pattern between the outer edge of the foot surface and the blade surface.*

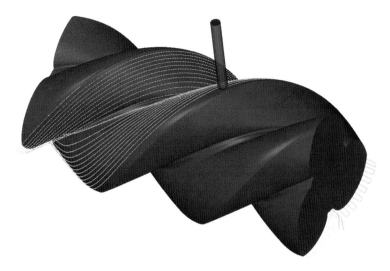

Figure 6-17 *Cut Pattern following the natural flow of the surface — the grid lines.*

Tool Axis Control

The examples shown in Figures 6-1 through 6-17 were designed to illustrate the results produced by tool motions on various parts. It is necessary to control the direction of the tool axis as the tool follows the **Cut Pattern**. The **Tool Axis Control** allows orientation of the tool's center axis to be manipulated as it follows the **Cut Pattern**. The sketches in Figures 6-18 through 6-25 illustrate these concepts

Figure 6-18 *The Tool Axis can be locked normal to a plane. In this example, the Tool Axis will be maintained normal to the bottom floor surface of each individual insert pocket.*

Figure 6-19 *The Tool Axis can be locked so that it always intersects any defined point on the holder side.*

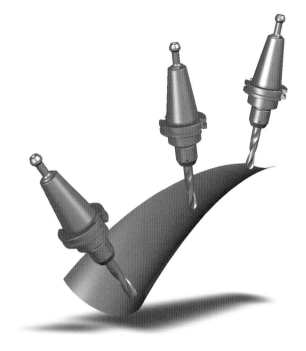

Figure 6-20 *The Tool Axis can be locked so that it is always aligned with a defined point at any distance as it follows the Cut Pattern.*

Figure 6-21 *The Tool Axis can be forced to remain normal to one surface or to multiple surfaces.*

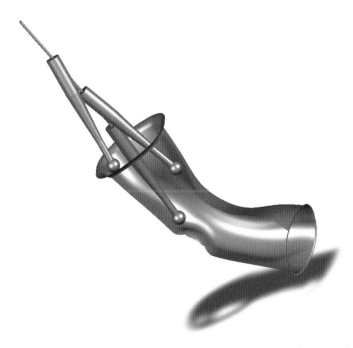

Figure 6-22 *The Tool Axis can be forced to follow a chain, while spiraling down an intake or exhaust channel.*

Figure 6-23 *The Tool Axis is controlled by the curves of the top and bottom surface edges.*

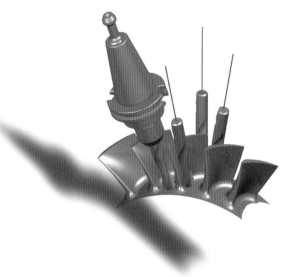

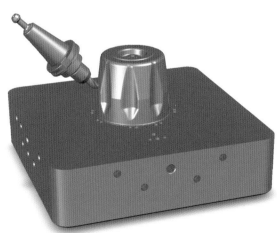

Figure 6-25 *A Tool Axis can be forced to rotate about any other axis.*

Figure 6-24 *Lines can be drawn that will guide the Tool Axis as it follows a Cut Pattern.*

In addition to the previously described **Tool Axis Control Methods**, more controls are available that allow the tool to be rotated around its tip by specifying lead, lag, and side tilt angles, as shown in Figures 6-26 through 6-30.

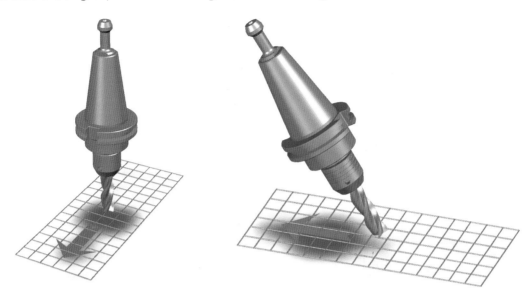

Figure 6-26 *Tool axis normal to a surface.*

Figure 6-27 Tool axis at a lead angle.

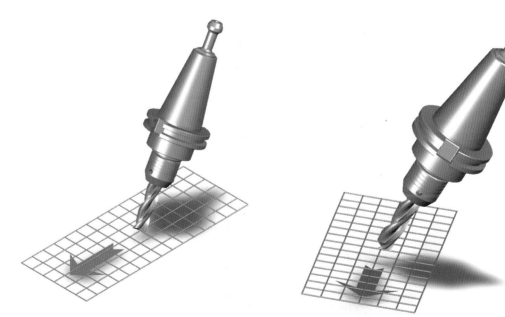

Figure 6-28 Tool axis at a lag angle.

Figure 6-29 Tool with side-tilt angle.

Newer systems even allow dynamic changes to be made to the side tilt, or the lead/lag angles, while cutting. The example in Figure 6-30 shows turbine blade machining in which the **Tool Axis** is dynamically controlled. With this control, the tool can be provided with optimum access to all the features on the blade in all stages of the cut.

Figure 6-30 *Dynamic side-tilt angle changes.*

Tool Tip Control

In summary, when CAD/CAM systems create 5-axis toolpaths, they will:

- First generate a number of tool positions along the user's chosen **Cut Pattern** as shown in Figure 6-31.

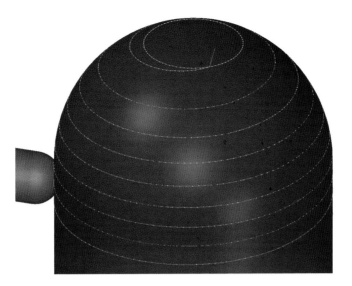

Figure 6-31 *Generating tool positions on the cut pattern.*

- The systems then assign tool vectors to every one of those positions, based on the **Tool Axis Control** method chosen by the user, as shown in Figure 6-32.

Figure 6-32 *The generated tool axis vectors.*

- Next, they will move the tool to a desired depth along the Tool Axis, based on the Tip Compensation method.

For example, surfaces generated to control a toolpath for the human head shown in Figure 6-33.

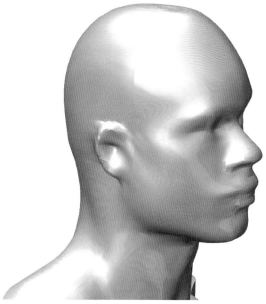

Figure 6-33 *Human head sculpted under computer numerical control.*

The surfaces were generated by a scanner, and therefore they are not the best quality. The file may have gone through a few translations. The model may have been scanned initially and saved as an IGES file, then sent to someone who saved it as a STEP file. Next, it could have gone to another shop where it was saved again as an IGES file. Every time a file gets translated between different CAD/CAM systems there is a tolerance issue. It is very easy for errors to be compounded and produce a poor quality CAD model. The model may consist of thousands of surfaces and there may be gaps between them. The **Tool Axis** would flip radically if it tried to stay normal to all the surfaces as it traveled. Fixing the gaps would be very time-consuming. A good, clean model will always produce a nicer **Cut Pattern**, stable tool axis orientation, and cleaner cuts.

A handy 5-axis trick is to create a clean core under the poor quality surfaces. This clean core is used to generate both the **Cut Pattern** and the **Tool Axis Control**. Then, compensation is applied to the tool tip in cutting the outer-skin surfaces, following the clean pattern.

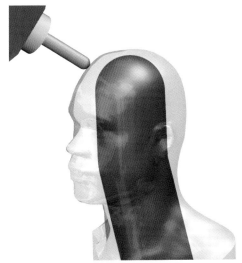

Figure 6-34 *A clean core was created under the poor-quality surfaces and the tool was moved to positions at the set depth.*

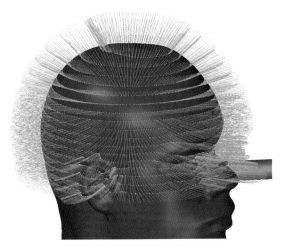

Figure 6-35 *The clean core was used to generate the Cut Pattern.*

Collision Control

It is a given that collisions or gouges are always to be avoided, so why is collision control needed? Why aren't all CAD/CAM Systems designed to avoid them automatically?

That first sentence above is not always true. In some instances, there is a need to gouge the drive surface! When would this application be useful? Engine head-porting is a good example. The shapes of the intake and exhaust ports are very complex. Traditionally these shapes were carved by hand, with carving tools similar to the instruments used by dentists. Reproducing these complex shapes has always been a challenge.

The CNC process is very good at reproducing shapes and comes in handy for this application. The challenge is getting these hand-carved shapes into the CAD/CAM system. Probing is a common method used to reproduce ports. A probe is a spherical instrument that is used to touch the part and record a point in space. Touching many points will record what is known as a point cloud which is a group of points that roughly represents the part's shape. If a probe of the same diameter as the tool to be used is employed, the tool can be guided along the points in this point cloud to cut the part. An example of a shape to be reproduced is shown in Figure. 6-36, and a close-up of the probe in contact with the surfaces in Figure 6-37.

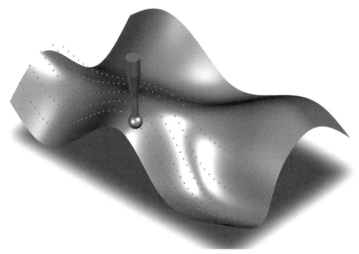

Figure 6-36 *Probe being used to generate points over the part's surfaces.*

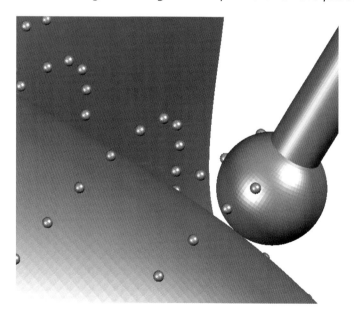

Figure 6-37 *Close-up showing contact of the probe with the part's surface.*

In some instances, it may be advisable to use a tighter cutting grid to obtain a better finish, or to use a different size of tool. In these conditions, it may be necessary to transform the **point clouds** into workable surfaces. These surfaces would exist relative to the center of the probe and, in this situation, it would be necessary to lead the tool center on the surface, (the same place where the probe center was) thus gouging the surface.

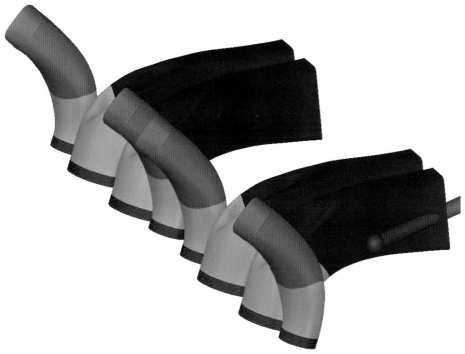

Figure 6-38 *These port surfaces were generated on the probe's centerline. The tool center is led on to the surfaces.*

Most engine builders today use either a more sophisticated scanning method that compensates automatically for the probe diameter, or laser scanners, as shown in Figure 6-39, that read the exact shapes of the ports.

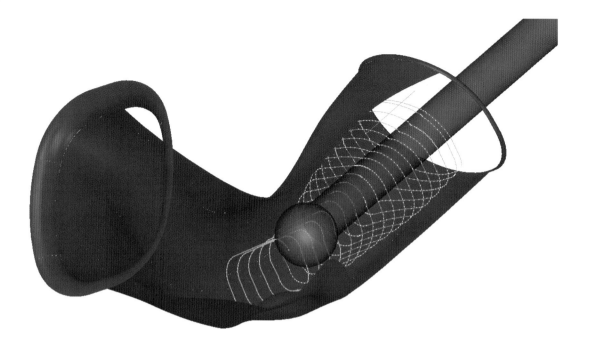

Figure 6-39 *This probe surface was generated with a laser scanner that can represent the true shape of the port.*

Collision avoidance must be used when cutting these complex surfaces. Collision control permits monitoring of the cutter's engagement with the surface, while ensuring that none of the other features of the tool (shank, holder, etc.) come in contact with any surfaces. Better CAM systems allow a choice of ways to avoid collisions, and even permit "near-miss" distances to be set for different parts of the tool.

The impeller example shown in Figure 6-40 has twisted and warped blades, which would be impossible to cut with the side of a tool. These shapes need to be generated by stepping down on each individual blade with a ball-nose cutter. The bottom fillet is small compared to the blade height, and although a long and skinny ball nose cutter is needed, it is not practical. A tapered-shank ball-nose cutter is preferred. Because there is very little room between the blades, there is great danger of gouging, both the blade being cut, and the neighboring blade. Caution is also needed at the hub surface to ensure that it doesn't get violated by the nose of the cutter.

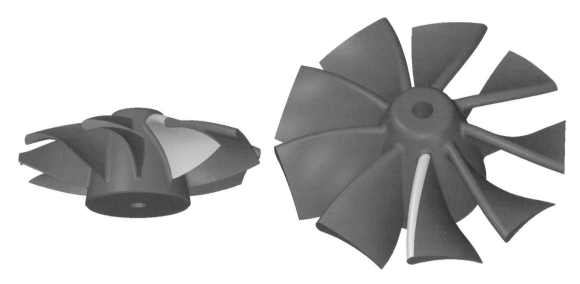

Figure 6-40 *A warped impeller.*

Some CAD/CAM systems provide control by allowing multiple avoidance strategies to be specified in the same path. For instance, in the above example it is possible to:

- Specify cutting with multiple, spiraling cuts.

- Specify that cuts should start from the top and work down toward the bottom of each blade.

- Specify use of a tapered-shank ball-nose cutter.

- Specify the side tilt angle that is to be maintained.

- If the cutter's shank comes within a certain distance from the blade, the tool is instructed to tilt away, either in the lead/lag, or the side tilt directions.

- If the tool nose comes in contact with (or within a near-miss distance) of the hub surface, it is instructed to retract along the tool's axis.

- If the tool holder comes within a near-miss distance from the top surfaces of the blades, the machine is stopped so that the tool can be moved out from the holder (longer tool is needed).

This level of control allows creation of a clean, smooth cut with a rigid tapered-shank ball-nose cutter, as shown in Figure 6-41.

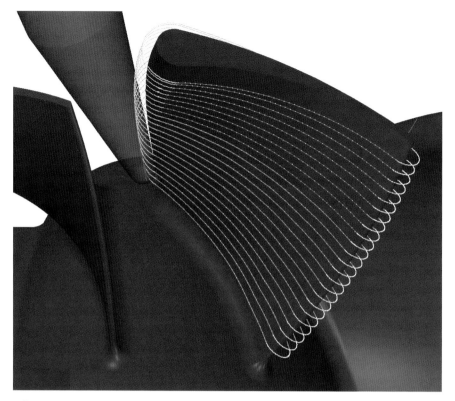

Figure 6-41 *A Clean Cut Pattern with dynamic tool axis control.*

Not all CAD/CAM systems provide this amount of control. Some will only allow the definition of check surfaces to be avoided, but will not provide the means to avoid them. Keep in mind that these controls focus on collisions between tools, holders, fixturing, and work-pieces. They will not avoid potential collisions on the machine. To avoid collisions between machine components, like rotary heads or tables, machine simulation is needed. That subject will be covered in the next chapter.

Additional Multiaxis Issues and Controls

Dovetail Effect

Even 4-axis, and especially 5-axis, motion will introduce some unique challenges. For example, if a straight tool is plunged into the center line of a cylinder and then the cylinder is rotated, a dovetail shape will be left between the start and end positions of the tool, as shown in Figure 6-42.

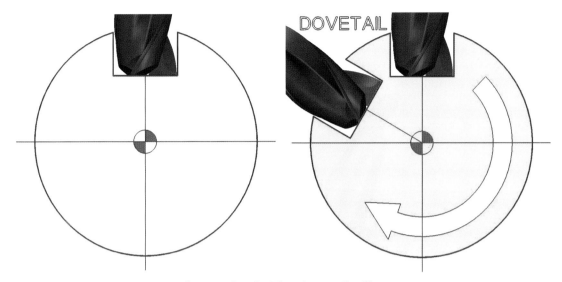

Figure 6-42 *The dovetail effect.*

If the intention is to cut a spline with parallel walls, the tool should be moved off center, as illustrated in Figure 6-43.

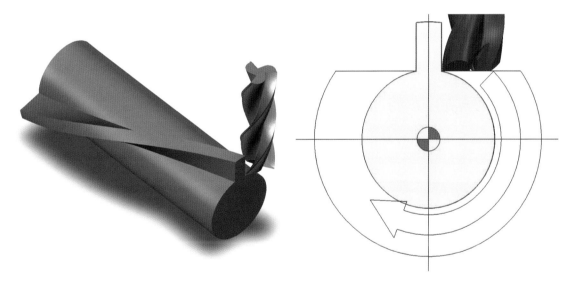

Figure 6-43 *For cutting a spiral spline, the tool must be moved off center.*

The offset amount must change for each side of the spline, and the offset amount will depend on the pitch of the spline. Note also that the bottom center of the tool face cannot be in contact with the minor diameter.

Cutting Direction

Most cutters are very sensitive to the cutting direction. In the 3-axis world, it is easy to see and define cuts that are conventional or climbing, but this is not true when cutting a multiaxis part.

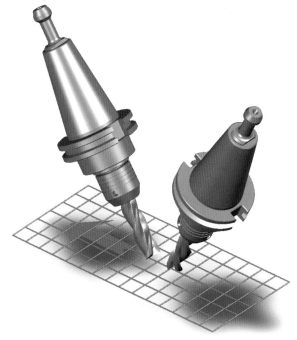

Figure 6-44 *Illustration of Lead Lag in milling operations.*

When taking a light cut, simply changing the lead/lag angle of the tool changes the cutting direction at the tool contact point, as illustrated in Figure 6-44.

The tool engagement area also changes drastically in deep or heavy cuts, such as those using lead and lag cut engagement shown in Figures 6-45 and 6-46.

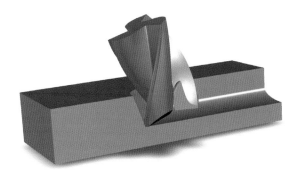

Figure 6-45 *Lead cut engagement in milling.*

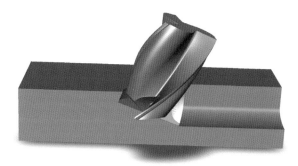

Figure 6-46 *Lag cut engagement in milling.*

The examples in Figures 6-45 and 6-46 show different engagements during the same cut, but changing the lead angle to a lag angle. The tool contact area changes from the side to the bottom of the tool. Extra attention must be paid to this aspect, especially if inserted, hollow-center, non-bottom cutting tools are in use.

Multiaxis Roughing

There are many instances where it is necessary to use long tools for roughing, as seen in Figure 6-47. This is usually dictated by the part features. Impellers are a good example of this problem. Tall blades with small gaps between them force the use of a long cutter, and these cutters don't perform well with side-cutting pressures. As the side-load increases, these tools will deflect, causing vibration, chatter, poor surface finish, and drastically shorter tool life. Multiaxis plunge roughing is a good way to remove material in these circumstances.

Figure 6-47 *Plunge roughing*

Plunge cuts should not be made to the final depth all at once. Instead, it is best to plunge only to a manageable depth, plunge out one layer, then pick away on the next one. The cutting pressure will be along the tool axis. This procedure will eliminate tool deflection and all its negative side effects. A typical job produced with this procedure is shown in Figure 6-48.

Some CAD/CAM systems also have the ability to look at the shape of the stock model and eliminate all air-cuts from the toolpath. This ability, combined with plunge roughing, can shave off hours from high-volume roughing operations. Plunge roughing is not a simultaneous multiaxis cut and therefore is a more rigid cut.

Figure 6-48 *This part was cut out of "green ceramic" which gets fired after milling. The finished component is resistant to abrasive chemicals in high-temperature environments.*

Machine Simulation

Machine simulation is the safest and most cost-effective way to prove out multiaxis toolpaths. Using a multiaxis machine to prove out programs is time-consuming and dangerous, both for the machine and for the operator! Running programs blindly on a real machine, based on a wireframe backplot in a CAD/CAM system, is just as dangerous.

When CAD/CAM programmers converse about programming a multiaxis machine, they typically use a special sign language involving rotating arms and torsos, while holding up two fingers and a thumb, signifying the right-hand coordinate system in all kinds of different orientations. They visualize the part and the machine as it performs an imaginary choreography. This visualization is not easy to do, especially if there are many different machine types in the shop.

Wireframe backplots portray the tool motion as it moves around a stationary part. This movement is later post-processed into machine motion and is different for every different type of machine. The CD included with this book contains a number of examples showing the same part being cut on various machines. It will be clear that even though the CAD/CAM backplot motions are the same, the machine motions are completely different.

With machine simulation, a machine's virtual replica can be shown on the computer screen where the cutting process can be simulated safely. This try-out will ensure that the program contains the most effective cut, the part is located in the machine's "sweet spot", and no fixtures, tools, or any machine components will meet unexpectedly.

It must not be assumed that machine simulation is only to be used for prove-outs with the sole aim of finding errors in the code. Instead, it must be looked at as an additional tool to help make clean, efficient, and accurate programs every time. With simulation, different approaches and different cutting strategies can be tested on different machines, without leaving the desk. And there is no need to tie down a machine for prove-outs. Nobody likes to see an expensive 5-axis machine sitting idle while programs are being tested.

People make mistakes under pressure. Even small mistakes on multiaxis equipment can quickly add up to catastrophic proportions. Damage to the part, machine, down-time for repairs, repair costs, and penalties can really ruin a business. Running a new, unproven, 5-axis program blindly on a machine, is like playing Russian roulette with the gun chambers fully loaded. Using multiaxis CNC equipment as a verification system doesn't make sense, and is much more expensive than using simulation. But with that said, nothing can substitute for the real thing. Even after simulation tests, the first run will always be exciting. The sights, the sounds, the feel of the cuts are irreplaceable. Machine simulation is not a magic bullet, but used properly, it is an extremely helpful tool.

Old School Simulation

Many shops still cut foam or wax for prove-outs. Some will even replace the cutting tool with a flexible pipe cleaner, and run the program on a finished part to see if there are any interferences. They will slow down, override both the rapid movements and the feedrates, and keep a close eye on all possible collisions. If there are any close calls they will stop, make changes in the program, either manually or in a CAD/CAM, and repeat the process. On a complex part, this process can take days. On a complex, state-of-the-art, multiaxis machine, this prove-out process could cost thousands of dollars in downtime alone. Only a highly-qualified operator/programmer should attempt this type of prove-out, even if it costs more in wages.

Realities

Even with today's advances in CAD/CAM capability, many people still manually edit the code created by their CAM system. There are various reasons for this and some of those reasons include the following:

- **Post processors are not configured properly.** For example, during rotary positioning moves, the rotary brakes should be disengaged, and engaged again during cutting. This brake application is governed by M codes that vary with different machines. If the post processor is not configured properly, these M codes will need to be inserted manually.

- **A repeating pattern on the part can be called up, using subroutines.** For example, an impeller has repeating features. Instead of letting the CAD/CAM write long extensive code, it is sometimes easier to take the CAM-created code for one feature and repeat it using subroutine logic. This procedure is particularly useful when there is a lack of memory in the machine's controller. No matter what the reason is for using this method, you will find that this more efficient program is always easier to prove out.

- **Manually-programmed probing routines are introduced.** For example, a branching/looping probing routine using system or user-defined variables might check the part for alignment at the beginning, or between tool changes. Then, based on the results, the probing routine would adjust the NC code to align with the part.

Experienced programmers tend to do more G-code editing than new programmers. New programmers tend to embrace and trust the technology more, and many are unfamiliar with G-code languages. As was established in earlier chapters, CAM systems first generate generic intermediate code (APT, NCI, CLS) and then post process that code into the machines' specific G-code language. All NC machines understand G-code, and when they read that code, they translate it into machine motions. Every word in that code, regardless of where it comes from — the CAM system or manual editing — will be recognized without discrimination. The most common question is whether to simulate the intermediate code or the G-code.

G-code Simulation Versus CAM Simulation

Only a handful of CAM systems have integrated machine simulation. Most of those only simulate posted toolpath code (XYZABC), not the posted G-code. Some have post processors that will post two streams of code at once — a simplified one for simulation and the controlling G-code for the machine. If these post processors are configured correctly, the virtual machine and the real one will behave exactly the same.

There is currently only one machine simulation software program that can run true G-code and that is Vericut® by CGTech. This program has multiple machine controllers available, and can be configured to realistically simulate all known G-code languages, including looping/branching logic, probing routines, and G and M codes. If configured properly, the program's virtual machines will behave exactly like the real ones. Note that both methods will only work if they are properly configured.

If the shop is programming manually, or does massive edits to the posted code, it will need simulation that is properly configured to simulate real G-code. On the other hand, if the CAM's post processor is properly configured to drive an on-board simulation, no other simulation is needed.

In either instance, the question to ask is "Who will do this configuration?" Configuring multiaxis machine simulation requires an intimate knowledge of each machine, the simulation software, and the post processor.

Configuring Virtual Machines For Simulation

Software companies have teams of dedicated professionals who spend all their time testing and applying the software. Every CAM developer has a Post department which writes translators (post processors) for every machines' language. The department is constantly monitoring new developments in the machine-building industry and is in close contact with the machine builder's applications teams. Together the teams develop factory-approved post processors. Without the efforts of the post writers, all CAM software would be useless. The ultimate end product of CAM software is not creating great toolpaths on a computer screen, but creating code that will govern the movements on specific CNC machines.

If true G-code machine simulation is the goal, Vericut by CGTech, is the best solution because their applications team has hundreds of years of combined hands-on G-code experience and is capable of configuring any type of CNC machine, even entire machining cells. The company specializes in reverse-post processing, meaning that they start with G-code and convert it to machine movements, just like a machine's CNC controller would.

Some CAM software packages offer multiple machine simulation interfaces. A few have direct interfaces with Vericut. Another popular choice is MachSim by ModuleWorks. The equally-capable ModuleWorks team specializes in post processors configured to produce both the simulation and the G-code output.

Every CAM and simulation software company provides post processing training and/or virtual machine building. These courses are typically a few days long. Companies can opt to send employees to one of those courses or just leave the configuration work to the professionals.

The following is an overview of the general steps in virtual machine building.

Virtual Machine Building

It is not necessary to virtually build an entire machine including the chip conveyor, NC controller, coolant tank, and so on. Such a process makes for slick simulation, but the only crucial part that needs to exactly resemble the real machine is the area near the working envelope. These motions must exactly replicate the real machine. The remainder of this chapter will cover the process involved to virtually build all the major machines that were covered in Chapter 2. The steps are very similar, regardless of the simulation software being used.

The Skeleton

The first step is to build the skeleton of the machine. The skeleton, or kinematic structure of the machine, describes how the machine's linear and rotary/pivoting axes are connected. Every machine will have a **BASE**, **TOOL**, and **STOCK** component. The best way to see the skeleton of the machine is to stand by the machine and jog every axis. Try to imagine the machine naked, without the covers. Observe the example in Figure 7-1.

Every machine will have a **BASE**, **TOOL**, and **STOCK** component.

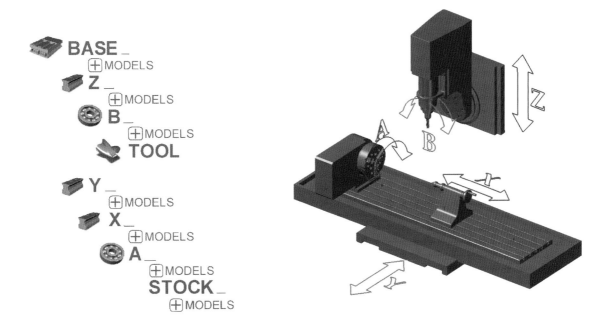

Figure 7-1 *Kinematic component tree.*

The base of the machine in Figure 7-1 is hidden, to allow a better view of the "business-end" of the machine. The kinematic component tree (shown to the left in Figure 7-1) describes the machine (shown to the right in Figure 7-1). The **BASE** is the first component. The **Z**-linear axis is attached to the BASE. The **B**-rotary axis is connected to **Z**. The indentation signifies the axis priority, meaning that if you move the **Z**-axis, the **B**-axis will move with it, because it is carried by the same slide. The last component on this branch is the **TOOL**, carried by, or attached to, the **B**-axis.

The second branch is also attached to the **BASE**, starting with the **Y**-linear axis component. Observe that **Y** is at the same indentation as **Z**. The **Y**-axis is carrying the **X**-linear axis component. **X** is carrying the **A**-rotary axis component, which in turn is carrying the **STOCK**, or workpiece. This kinematic component tree is the most basic description of a machine, and is a stripped-down skeleton of the machine. There are no models attached to this skeleton, but you can tell by a glance which bones are connected together.

Many other component types can be attached to this basic structure including, fixture, tool changer, pallet changer, and robots.

Components vs Models

Depending on which simulation software is in use, multiple models can be attached to every one of the main components. This ability to attach models enables different properties to be assigned to each of the models. Unique tolerance values, colors, translucency, visibility, and reflectivity can be assigned to each model, and individual models can also be included or excluded on the collision-detection settings.

Most machine simulation software uses STL models as a default, and some can also use solid primitives (block, cylinder, cone, sphere, or torus). Other software can use its native solid models, or a mixture of all the above models.

Some popular machine examples are illustrated in Figures 7-2 through 7-10.

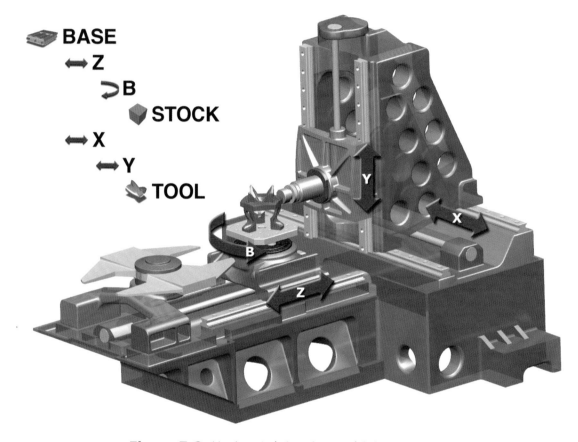

Figure 7-2 *Horizontal 4-axis machining center.*

The horizontal 4-axis machining center configuration shown in Figure 7-2 is very popular for high-volume tombstone-fixture type manufacturing. Note the pallet changer, which can be adapted to service an entire pallet center. With this capability, multiple different jobs can be introduced into the manufacturing process without stopping the machine.

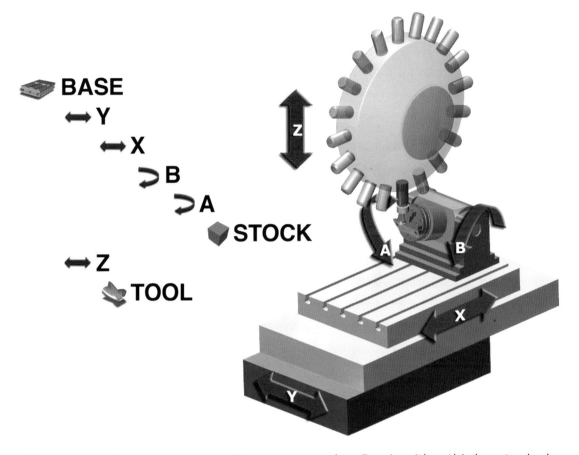

Figure 7-3 *Vertical 3-axis machine, converted to 5-axis with a third-party dual rotary device.*

The modifications shown in Figure 7-3 can be adapted to suit most 3-axis vertical machining centers. The dual rotary device bolts to the machine's table, instantly transforming it into a 5-axis machine. Some room will be lost in the Z-axis working envelope, but the multiaxis capability will be gained.

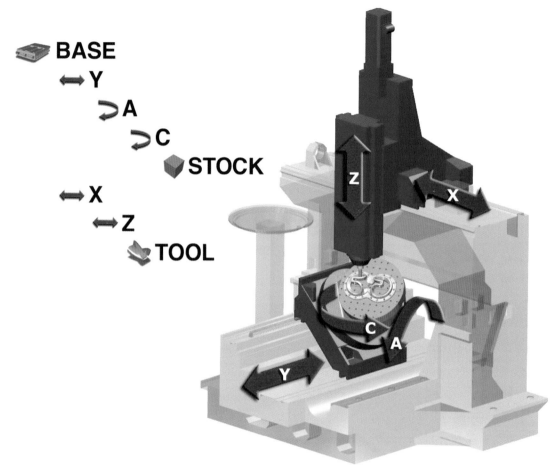

Figure 7-4 *Vertical 5-axis machine with a dual, rotary, nutating table.*

The machine in Figure 7-4 is a dedicated 5-axis **Table/Table** vertical machining center. Note the rigid machine base. Such a machine can handle heavy work with precision and confidence.

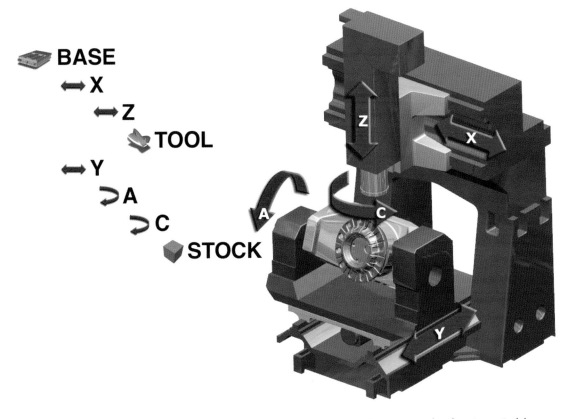

Figure 7-5 *Vertical 5-axis machine with a trunnion-type dual rotary table.*

Trunnion-type dual rotary configurations, as shown in Figure 7-5, are very popular in the industry. This may be because they are competitively priced and easy to set up and operate.

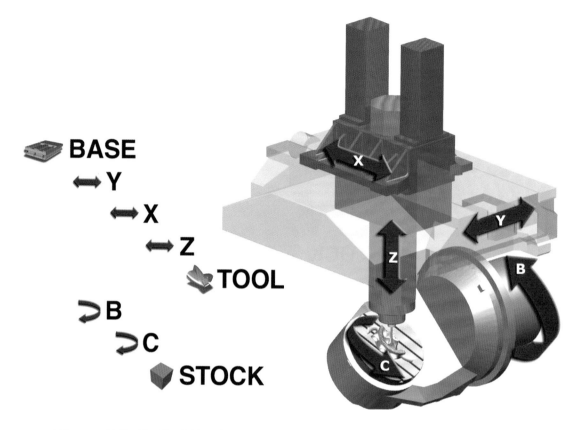

BASE
⟷ Y
⟷ X
⟷ Z
TOOL
⟳ B
⟳ C
STOCK

Figure 7-6 *Vertical 5-axis machine with a dedicated dual rotary table.*

Figure 7-6 shows another example of a sturdy, dual-rotary, 5-axis vertical machining center. This machine also has the ability to spin the C-axis as a spindle, allowing for turning work to be done.

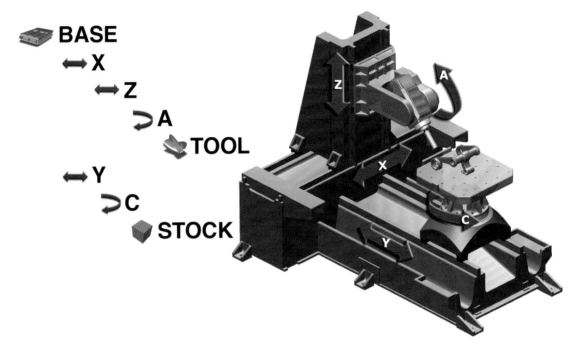

Figure 7-7 *Horizontal/vertical 5-axis Head/Table machining center.*

The machine in Figure 7-7 is called VH — Vertical and Horizontal. It is a 5-axis **Head/Table** machine, and its design allows for exceptional flexibility in addition to formidable rigidity.

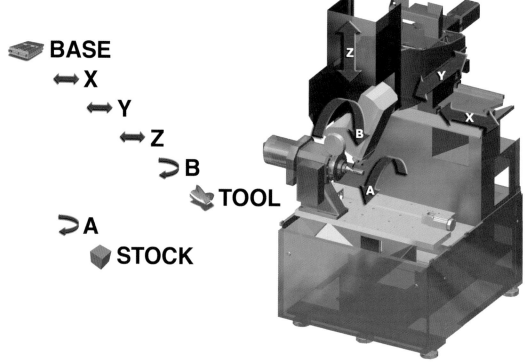

BASE
⟷X
⟷Y
⟷Z
⤵B
TOOL
⤵A
STOCK

Figure 7-8 *Vertical 5-axis Head/Table machining center.*

The vertical 5-axis, **Head/Table** machine in Figure 7-8 provides an amazing combination of speed and precision.

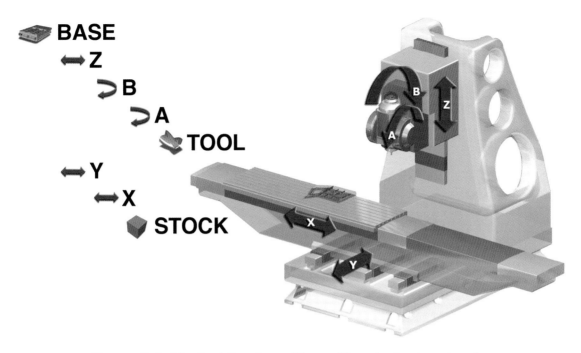

Figure 7-9 *Vertical 5-axis profiler, with a dual rotary head.*

Many manufacturers offer variations on the type of **Head/Head** configuration shown in Figure 7-9, commonly known as a profiler. Typically these machines have limited rotary range combined with long bed travel.

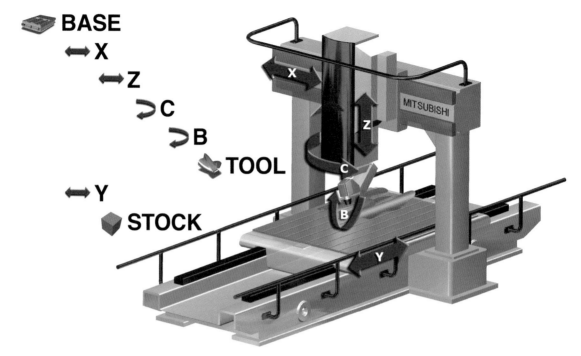

Figure 7-10 *Vertical 5-axis laser machine, with a dual rotary head.*

The vertical 5-axis machine shown in Figure 7-10 is used for laser-machining, but this kind of **Head/Head** configuration is also very popular for milling and water-jet machining.

Machine Simulation Interfaces

A GUI (Graphical User Interface), or form of text file, can be used to build virtual machines. With such a program, models, or whole component branches, can be manipulated individually. For example, the virtual machine can be used to translate, rotate, or set dependencies, translucencies, or reflectivity.

Once the virtual machine is built, all its axes can be moved individually with MDI (Manual Data Input) commands, or slider bars, to check if the correct models are assigned to the correct axes. These commands can also be used to check if the positive and negative motions are correct. Remember that all simulation software is useless if it is not emulating the movements of the real machine. The models representing the real machine must be accurate in relation to the business end of the machine. This area is near the work envelope and includes the spindle, fixturing, and rotary devices.

Once the physical model of the machine is built, the virtual controller must be configured. In a CAM system this work is done with the post processor. In Vericut, configuration is achieved with a reverse post processor. This configuration process is critical in emulating the behavior of the real machines.

Using Machine Simulation

These days, very few people program exclusively by hand. Most people use a CAD/CAM system to generate code. The part is typically either designed or imported, and then toolpaths are generated using tools from an internal or an external library. Machine simulation can be run at any time during or at the end of this process, provided the groundwork has been laid down and the machines have been built.

The process of setting up machine simulation is very similar to setting up a real machine. The part must be placed on the machine in the correct orientation and then the **Local Coordinate System** needs to be set relative to the **Machine Rotary Zero Position**. The tools then need to be loaded into the magazine and the **Tool Length Offsets** must be set correctly. This work can be time-consuming if there is no direct interface between the CAD/CAM and the simulation programs. If there is a well-configured interface, or if the simulation is an intricate part of the CAD/CAM, then setting up will take only a few seconds of processing time.

Native CAD/CAM simulation loads tools from its libraries. Vericut uses its own tool manager, or it will build a tool library automatically if it is integrated with a CAM system. Once the part, tools, and toolpaths are loaded, the simulation is ready to be run, either as single blocks, or continuously. The simulation can be slowed down or sped, and the model can be dynamically rotated. Some systems allow movements forward or backward at any time, but others don't offer this option. Some systems will show material removal with simulation, and some will permit analysis and measurement of the virtual part. Most systems will signal if there is a near-miss or collision between any configured components. They will also display an alarm if the limit switches are hit by over-travelling on any of the motion axes. Operators are able to see through models by making them invisible, which allows examination of the cutting process in ways that are not possible on a real machine.

There are many benefits to machine simulation, which allows different ideas to be tested out without pressure. Estimated program cycle times can be accessed, to help determine the best one. Crashing a machine on the computer screen is not a big concern, whereas crashing a real one is a catastrophe. But not using a multiaxis machine to its full potential is a shame. Simulation allows the best ideas from different cutting strategies, and the most efficient motion for any specific machine to be combined.

The process of setting up machine simulation is very similar to setting up a real machine. The part must be placed on the machine in the correct orientation and then the **Local Coordinate System** needs to be set relative to the **Machine Rotary Zero Position**.

NOTES:

8

Selecting The Right Machine
For Your Application

Making a multiaxis equipment choice decision is similar to choosing a car make and model. The decision needs to be based on the intended use, budget, and personality, along with many other considerations. The multiaxis "garage" includes the equivalents of race cars, all-terrain vehicles, buses, and luxury vehicles. There are general-purpose machines and there are machines made for specific applications. This chapter may help narrow the search based on the specific parts being manufactured.

Most small shops enter the multiaxis arena by adding a single- or dual-rotary unit to their existing 3-axis vertical machining center. The addition of the single- or dual-rotary unit allows parts to be manufactured more quickly and makes it possible to machine more complex parts that were previously out of reach. This advance may cause a chain reaction. When shops get better at producing complex parts, they start to charge more for those parts. They then seek out even more challenging work to make more money. In turn, these ventures will stretch the limits of capability of the equipment, prompting consideration of purchasing more new equipment.

The available budget is always the big consideration. The price of any machine will reflect its quality, but as with cars, the price may also be affected by the name brand. However, budgetary considerations are outside the scope of this book.

Machine manufacturers spend a great deal of time developing machines. They also spend time on their sales and marketing efforts. Reputable manufacturers have applications teams who install new equipment, train new customers, and provide ongoing technical support. They also employ dedicated applications specialists who can prepare benchmarks, or turnkey solutions, for prospects and customers.

Regardless of the specific machine type under consideration, it is smart to research the reputation earned by the support services provided by different manufacturers. Most CNC equipment is sold by a dealer network. Not all dealers will maintain the same quality of service. It is wise to visit local shops that have different CNC equipment and talk to them about their experiences. Ask them how their equipment is performing, what the service is like when there is a problem, and if the manufacturer provided good training. It may also be wise to ask if the suppliers delivered on all their promises.

Select a machine manufacturer that suits the applications criteria, and then take a good look at the variety of parts currently being manufactured in your plant. Also consider the parts you intend to manufacture in the future. Consider the following scenarios.

How many parts are typically run after each set-up?

If your shop produces 500,000 of the same parts per year, it would be wise to look for a dedicated machine or machines to produce that part. Investigate the possibility of a turnkey solution from the machine builder. Such a solution may include a complete machining cell, possibly with multitasking machines and robotic loaders.

Does your shop/company thrive on challenging jobs and have a reputation for producing complex work?

Some shops like to take on work that others consider to be too difficult. These companies learn from every challenge and become better and better with every job. Taking on difficult jobs may be risky, but it can pay great dividends. Before contracting for such demanding jobs, ensure that your multiaxis equipment is flexible, precise, and adaptable enough for the challenge.

Are your existing CNC machines waiting for programs, or are your CNC programmers waiting for a free machine?

If existing equipment sits idle waiting for programs, then the workflow, CAD/CAM system capability, and programmers' and operators' proficiencies need to be scrutinized. If programmers are waiting for free machines, it is again a good idea to check the CAD/CAM system's capability. Could the cutting strategy be improved? Are the right tools being used? Imagine running old style high-speed steel tools on a modern CNC machine capable of 40,000 RPM and 1500 IPM — the limitations of cheap tooling could hold back a very capable and very expensive machine. In the same way, if your CAD/CAM system is obsolete, you won't be able to use your CNC equipment to its full potential.

Are you happy with the performance of your CAD/CAM system, and are you using it to its full potential?

Make sure that your CNC programmers are up-to-date with their training on your CAD/CAM system to ensure it is being used to its full potential. It is much cheaper and easier to get organized, trained, become efficient, and promote teamwork, than it is to buy a brand new machine and put it into production.

Is your shop/company dedicated to a single manufacturing field, for instance automotive, aerospace, mold & die, medical or oil?

The manufacturing field you are working in will also affect your choice of machine type. There are different torque, speed, and precision requirements in every field.

New Possibilities

After determining that your shop is running full out and needs additional equipment, it is time to consider new possibilities. The first obvious consideration is the physical size of the machine, and that is dictated simply by the size of the parts that will be machined and the size of your shop floor. The next consideration is the material that will be used, which will determine the rigidity needed. The quality requirements of the machine will be affected by the expected tolerances you want to hold, and budgetary restraints must also be kept in mind. Aside from these properties, keep in mind that some multiaxis equipment is better suited for certain types of work than others.

Head/Head Machines (with long X- or Y-axis linear travel, but limited rotary axis travel)

The manufacture of airplane wings and fuselage panels is a good fit for **Head/Head** machines. The panels are designed for strength, but are kept as light as possible. There are several tapered-wall pocketing machines that are perfectly suited for swarf-type toolpaths. Typically these parts are made from solid billets in two set-ups, as shown in Figure 8-1.

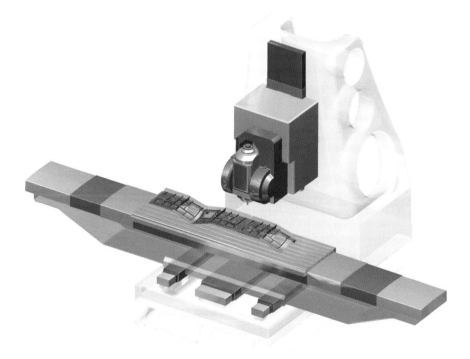

Figure 8-1 A vertical mill set-up for machining an aerospace panel.

An airplane wing stringer is a good example of a part that is long, but very slim. Parts like this are typically machined from special extrusions, which can be over 40 feet long. Typically, parts like these were made on machines similar to the ones shown in Figures 8-2 and 8-3, using multiple set-ups and elaborate fixturing.

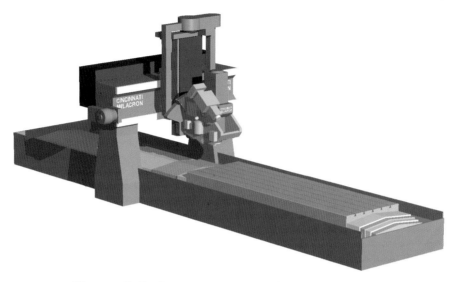

Figure 8-2 *Gantry-type Head/Head machine.*

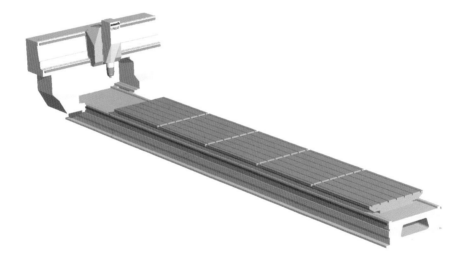

Figure 8-3 *Bridge-type Head/Head machine.*

The parts would tend to deform between set-ups because material would be removed unevenly, first from one side, then from the other, in a second setup. The machine shown in Figure 8-4 solves this problem.

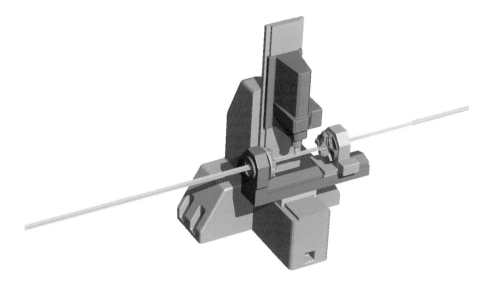

Figure 8-4 *Dedicated extrusion milling machine.*

The machine in Figure 8-4 is well suited for machining long extrusions. It is a 5-axis machine with X, U, Y, Z and A-axes. The U-axis moves parallel with the X-axis and it has two sets of rotary jaws that are used to clamp and traverse the extrusion past the cutting tool. Cutting takes place in a narrow but rigid corridor in successive sections. The overall lengths of the parts are limited only by the support systems at either side of the machine.

Head/Table Machines (*with long X-axis travel*)

Long parts, similar to the examples shown in Figure 8-5, require severe rotary motions in the primary axis and limited rotary motions in the secondary axis.

Figure 8-5 *Typical rotary parts.*

These parts would be best manufactured on the **Head/Table** machine configuration shown in Figure 8-6.

Figure 8-6 *Head/Table type milling machine.*

The rotary pivoting configuration shown in Figure 8-6 is very suitable for manufacturing long rotary parts. The weight of the part is supported by a tail stock, and the part is rotated around its center of mass. Inertia is an important consideration when using multiaxis machines. Consider the configuration for engine head porting shown in Figure 8-7, and imagine the differences in machine movements when compared with Figure 8-8.

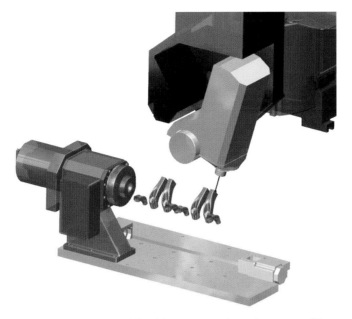

Figure 8-7 Head/Table engine head-port milling.

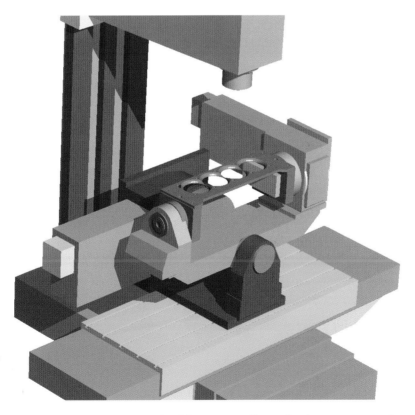

Figure 8-8 Dedicated Table/Table port milling dual rotary attachment.

The machine pictured in Figure 8-7 is designed to rotate the head around its center of mass without generating unwanted centrifugal forces. The machine in Figure 8-8 has something called a "Rock-and-Roll" dual-rotary device. It is designed especially for machining ports on engine heads. The entire fixture holding the part is rocked and rolled throughout the cutting process to present the work to the cutter. These fixtures need to be carefully balanced to ensure smooth motion.

Head/Table Machines

Head/Table configurations such as those shown in Figures 8-9, 8-10, and 8-11, are among the most versatile choices for a variety of other multiaxis applications. This versatility derives from the fact that the steady rest can easily be removed and the space can be used for mounting additional fixtures. Customized fixtures can also be built to suit special jobs.

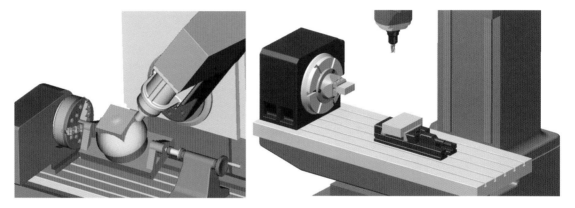

Figures 8-9.and 8-10 *Additional versatility using multiple fixtures.*

Figure 8-11 *Machining an auger feed spiral for an injection molding machine.*

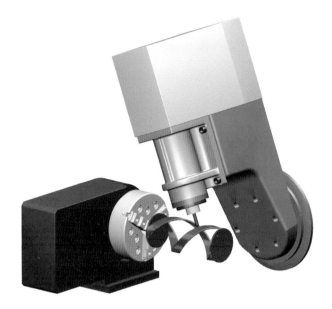

Figure 8-12 *Machining a rotary windmill unit.*

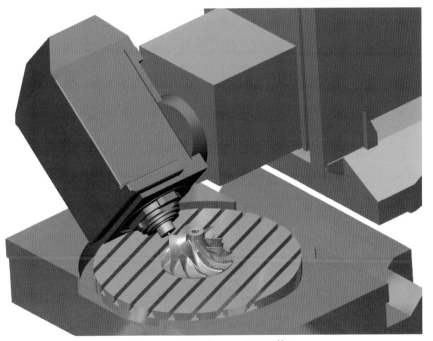

Figure 8-13 *An impeller.*

Figures 8-12 and 8-13 represent examples of vertical machines with long X-axis travels, but **Head/Table** machines are built in many forms and shapes.

Rotary Table— Tilting Head Combinations

The example shown in Figure 8-14 blurs the line somewhat between the vertical and horizontal definitions.

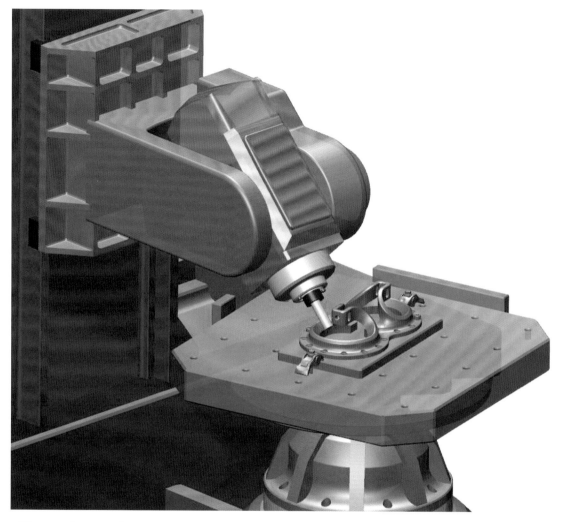

Figure 8-14 *This Head/Table machine is available in both vertical and horizontal configurations.*

The rotary-table and tilting-head configurations shown in Figures 8-15 through 8-18 are not suitable for long parts, but can readily be adapted for a variety of multiaxis applications.

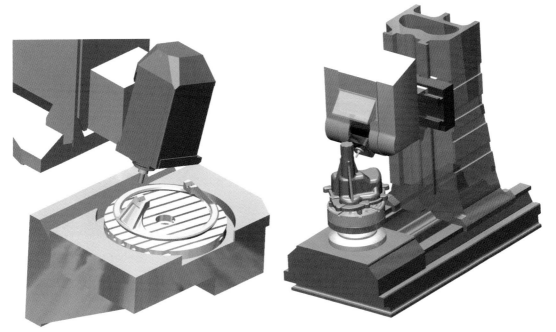

Figures 8-15 and 8-16 *Head/Table aerospace, and Head/Table automotive applications.*

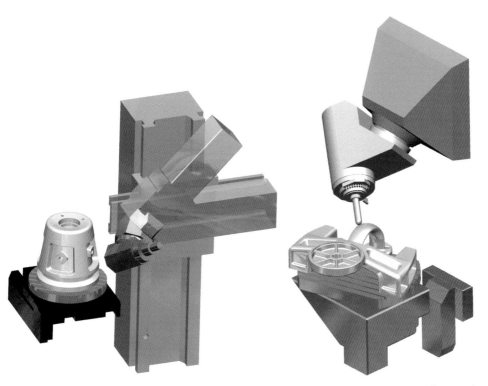

Figures 8-17 and 8-18 *Two Nutating Head/Table configurations.*

All rotary-table, tilting-head machines tend to rotate the workpieces around their centers of mass while maintaining the capability to reach all their features by tilting the head. These machines are built in many sizes and are widely used in many different industries, from manufacturing small medical parts (Figure 8-19) where precision and speed are the main requirements, to manufacturing large earth-moving equipment parts (Figure 8-20), where rigidity and horsepower are the focus.

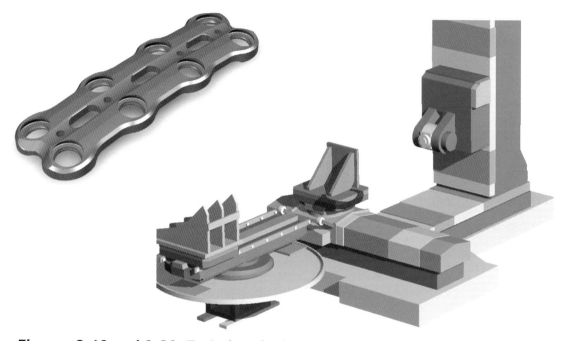

Figures 8-19 and 8-20 *Typical medical part, and heavy equipment component manufacturing.*

In the mold and die industry, most of the roughing operations are done on 3-axis, vertical or horizontal machining centers. In this manufacturing field, one of the challenges is cutting deep cavities or tall cores. The deep cavities are designed with steep side walls, usually at angles of 1 or 2 degrees, and often require uneven floors with small fillets along the intersection of the wall and floor surfaces, as shown in Figure 8-21. Cutting these fillets on a 3-axis machine would require long, ball-nose cutters. Small steps need to be taken, causing long cycle times. The tool is often deflected by the high cutting forces, causing vibration, excessive cutter wear, and poor surface finish.

Figure 8-21 *Typical plastics-mold cavity.*

Using a 5-axis machine allows tapered ball-nose cutters to be used for this work. The tapered configuration makes the ball-nose tool much more rigid for the same diameter, and the ability to tilt the tool also allows use of a shorter cutter, as shown in Figure 8-22. More aggressive cuts can then be taken, shortening the cycle time. Deflection of the rigid tool is less, and vibration is eliminated due to the reduced deflection. Tool life is increased, and a precise, good-quality surface finish is achieved.

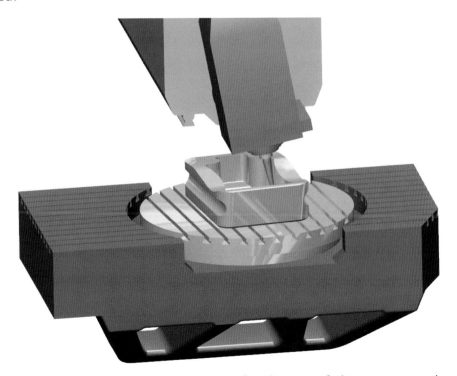

Figure 8-22 *Multiaxis machining allows for the use of shorter, tapered cutters.*

Table/Table Machines

Figures 8-23 and 8-24 show the most common configurations of **Table/Table** machines. The parts to be machined are clamped to a dual-rotary table and are rotated around the tool. Inertia is a consideration. The dual-rotary table is either mounted on the machine table or is a dedicated dual-rotary component of the machine. These machines are not suited for manufacturing long parts. The work envelope is fairly limited, especially when some tool changer limitations are considered. Despite the limitations, this configuration is very popular.

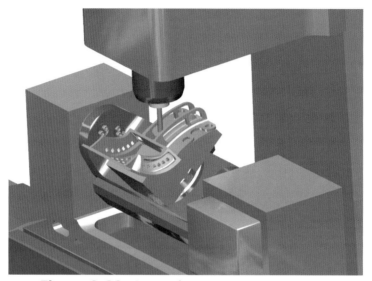

Figure 8-23 *A popular trunnion type setup.*

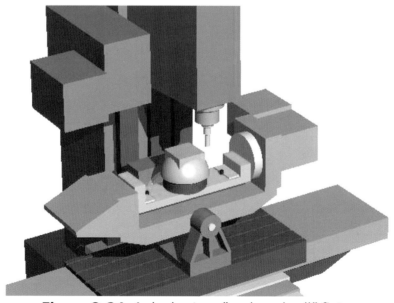

Figure 8-24 *A dual rotary "rock and roll" fixture.*

Table-mounted units are not completely rigid, but dedicated dual rotaries can be both agile and rigid. They are equally well suited for 3+2 indexing work, and for simultaneous multiaxis work. Some applications are shown in Figures 8-25 through 8-28.

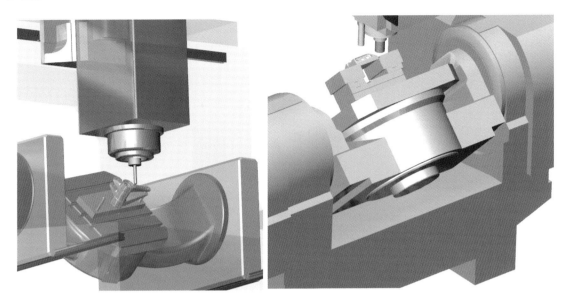

Figures 8-25 and 8-26 *Machining an aerospace bracket, and a fixture component.*

Figures 8-27 and 8-28 *Machining rotor blades, and machining a medical component.*

Gantry Type Head/Head Machines

Gantry type **Head/Head** machines, as shown in Figure 8-29, are used for large parts, mostly in the aerospace, oil, and wood industries. This configuration permits long linear travels. Some machines are designed to allow changes of heads in addition to tools. Rigidity and precision may not be the strong suit of these machines, but long reach capability is.

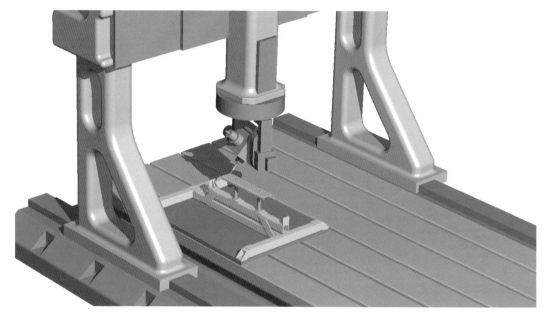

Figure 8-29 *Water-jet/milling combination machine.*

Some more machine variations are shown in Figures 8-30 through 8-33. However, it is impossible to describe all the different machine configurations that are available, especially because this is a constantly evolving field.

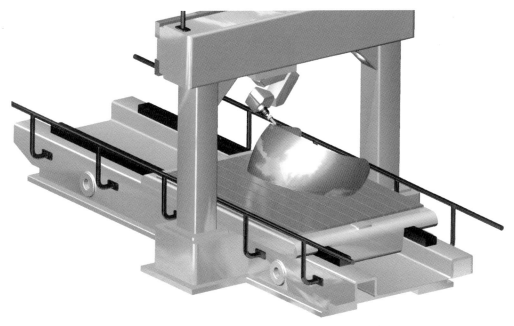

Figure 8-30 *A 5-axis laser cutting machine.*

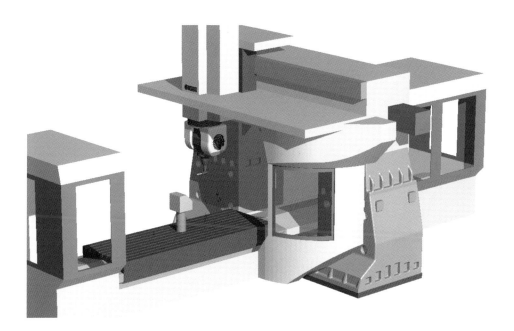

Figure 8-31 *This machine presents a good compromise between long reach and rigidity.*

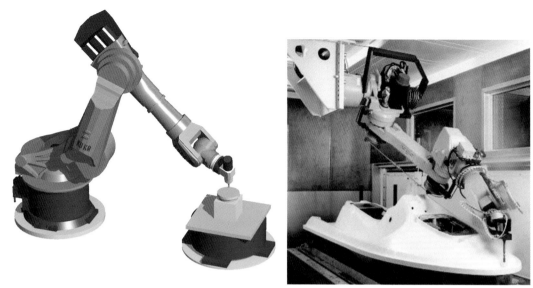

Figures 8-32 and 8-33 *A 6-axis industrial robot, and a 7-axis industrial robot.*

This chapter has only covered the most popular designs, and some suggested applications based on experience. It is recommended that engineers spend some time on initial research when choosing a machine, research not only of the machine, but also the intended use.

Choosing a CAD/CAM System For Your Application

Choosing the appropriate CAD/CAM system is as important, if not more important, than choosing the most suitable multiaxis machine. There are many specialized machines that are dedicated to specific types of work, however one CAD/CAM system will drive all the CNC equipment in the shop.

It is important to make sure that the selected system can handle not only all the different types of work the shop does now, but will also be capable of taking on future challenges.

CAD (Computer Aided Design)/CAM (Computer Aided Manufacturing) is always referred to as one combined system because most CAD/CAM systems offer both design and manufacturing capabilities. Be aware, however, that very few excel in both CAD and CAM.

Systems with heavy CAD emphasis have their roots in CAD and are better at solid modeling so that they can handle large assemblies with ease. These systems have associativity between all the components so that when a change is made to one feature on one part in an assembly, it will propagate throughout the entire assembly. These systems are very good at managing CAD data, but their CAM capability may have been added later and it often does not have the same depth.

Systems with heavy CAM emphasis are good at everything related to toolpath creation, from simple 2D drilling, contouring, and pocketing to multi-surface and multiaxis machining. Toolpaths can be generated for all kinds of CNC equipment including wire- and other- EDM, water-jets, lasers, lathes, mills, and multitasking machines. These systems have intelligent tool libraries with associated feeds and speeds for different materials and cutter types. Instead of heavy CAD capability, these systems are very good at importing CAD data from any system, with the main goal of generating a toolpath from that data.

Special Purpose Software

Many specialized CAD/CAM systems have been designed for specific purposes. For example, some shops in the mold and die industry use CAM systems that have virtually no CAD capability, but they can import large, complex, multisurface files quickly. The user only needs to choose the tools and select from a short list of automated cutting strategies. A toolpath is soon generated, posted, and ready-to-go. The trade-off for this speed and ease is realized when engineering changes are necessary. Those changes need to be made on a separate CAD system and imported back into the CAM software. Also, these specialized CAM systems will not support any other kind of CNC machines (lathe, EDM, plasma, waterjet, etc.) and many won't even support simple contour, drill, or pocket routines. This type of special purpose CAM software only makes sense for shops that are machining large mold cavities day in and day out. It may be necessary to purchase a separate programming seat of CAD, and maybe even another seat of general purpose CAM.

Software that can dynamically change the feedrate throughout the cutting process is another good example of a specialized CAM feature. This feature mimics an operator standing at the machine and overriding the programmed feedrates by manually manipulating the feedrate override dial. In mold and die manufacturing, large amounts of material need to be removed. The topography of multisurface molds is often so complex that it is impossible to maintain a constant step-over, or even a constant depth of cut. Cutting forces on the tool vary greatly throughout the process of machining large molds and dies, and the work can take hours, days, and even weeks. It would be impossible to stand by the machine and anticipate every motion of the axes, and override the corresponding feedrates, but with feedrate optimization, the software will vary the rate automatically. This optimization takes place before any cutting is done, based on constants for volume removal rate, chip-load, surface speed, and other factors. Feedrate optimization produces constant cutting forces that are designed to lengthen tool life, increase accuracy, and dramatically shorten cycle time.

Software that is specifically designed to generate toolpaths for lathes is one more example of specialized CAD/CAM. This software offers limited CAD capability and only turning-specific toolpaths. The software is often built in to the controllers on certain machines, and only generates toolpaths specific to that machine's conversational language. With this type of software there is no need for a post processor. The approach is very direct, and that can be an advantage. However, it can also be a disadvantage because these toolpaths cannot be transferred to any other machines. Grinders, lasers, water-jets, plasma cutters, and other specialized machines can all operate in this same fashion.

In addition to CAD/CAM systems, other tools are available that can close the loop between design and manufacturing. Simulation software packages can help check and optimize the results generated by CAM software, and are a very important link between the virtual and physical worlds. Ensuring that toolpaths are bullet proof in the virtual world will save the shop time and money in the long run. These tools should not be overlooked when the shop is being outfitted for multiaxis work.

CAD/CAM Toolbox

Buying a CAD/CAM system is like buying a fully stocked toolbox, but care must be taken to ensure that it contains the right tools for the job. All tradesmen have their own idea of the perfect set of tools. A perfect set of sharp, high-quality chisels would be useless to an electrician. At the same time, it would be cumbersome to use a Swiss Army knife as a screwdriver all day long.

Very few CAD/CAM systems can do everything well. They all have their strengths and weaknesses. On the other hand, very few companies need all the power afforded to them by a modern CAD/CAM software system. The trick is finding the right balance.

Some CAD/CAM companies provide for the capabilities of their software to be increased as the company grows and demands more functionality. Most first time CAD/CAM users will start off with software that can perform only simple 2D drilling, contouring, and pocketing toolpaths. Once the users become proficient, they can take on more complex, 3D, multi-surface machining, or multiaxis 3+2 indexing work. From that point, users can move into complex, simultaneous, multiaxis milling, or even operation of multi-tasking milling/turning machines.

Multiaxis CAD/CAM Considerations

Multiaxis manufacturing requires software that is very strong in CAM. CAD capability is needed, but mostly to import CAD files from all the major CAD systems, in all the popular CAD data formats. On top of that requirement, additional CAD capability is needed to create supporting geometry for tool axis control, fixture design, or virtual machine building.

High-end CAD/CAM systems are fully associative. If a design change is made, the change will propagate through the entire database and will modify the necessary movements in the toolpath. This feature is helpful if one software package is used for the entire design-to-manufacturing process. If a single, all-encompassing package is not used, then extra cost is being incurred for associativity that cannot be used. Unfortunately, most geometry associativity only works with native geometry.

Most multiaxis shops import files from a variety of customers. These files could have been designed in any number of CAD systems, so it is crucial to be able to read and write in multiple CAD/CAM languages. Once the model is imported, it is critical to have good analysis tools to analyze it and then separate its major features into organized layers or levels.

After the model has been analyzed and organized, some additional geometry creation may be needed. This geometry could include additional wireframe, edge curves, lines, arcs, points, non-trimming surfaces, or even some solid model creation. This work will require light-duty CAD capability.

Multiaxis CAM

The category of 3+2 indexing work requires the ability to quickly and easily change the work planes, which are always perpendicular to the spindle/tool axis. The creation and manipulation of these work planes, also known as **Active Coordinate Systems**, should be intuitive and easy-to-use. Some systems work interactively by allowing the user to simply pick a solid face, an arc, two lines, three points, and such. to define the orientation of a new **Active Coordinate System**. This selection is a light-duty capability for most CAM software.

Heavy-duty CAM capability is needed for tackling simultaneous multiaxis applications. This capability has to be a delicate balance between control, flexibility, and ease of use. A shotgun approach doesn't work well here — the precision of a rifle is needed.

Consider mold and die work as an example. This work is one of the most demanding and accurate fields in manufacturing. Molds cannot be mass-produced but are made one or two at a time, and they have predictable features, either a core, or a cavity, or a little of both. A good 3-axis roughing strategy will always work well here. Some CAM systems can quickly and automatically analyze the features and then automatically generate a toolpath to machine them. In this shotgun approach, a wide field of targets can be covered with one shot.

Precise control is needed when it comes to driving simultaneous multiaxis machines. The following is a list of must-have tools from a well-rounded, multiaxis, CAM software toolbox. Please refer to Chapter 6 for detailed examples.

- **Cut Pattern Control**
  It is important to have a variety of ways to define and control the pattern that will be followed by the cutting tool. These patterns can be anything from a simple wireframe to complex surface patterns such as that shown in Figure 9-1.

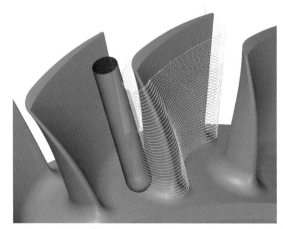

Figure 9-1 *Spiraling cut pattern on a turbine blade.*

- **Tool Axis Control**

  Tool axis control provides the ability to set and manipulate the center axis alignment of the tool during the cutting process, as illustrated in Figure 9-2. These controls can be dynamic or static, but it is essential that they work in a predictable, stable way.

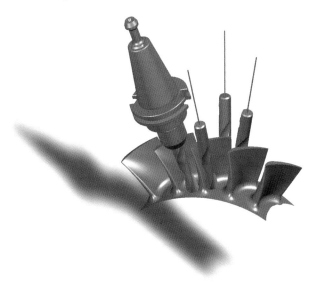

Figure 9-2 *Positions of tool axis controlled by lines.*

- **Tool Tip Control**

  The tool tip control targets the precise area of the tool tip's engagement with the part, as shown in Figure 9-3.

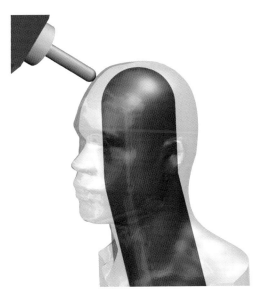

Figure 9-3 *Tool tip compensated to follow the outer surfaces of the work.*

- **Collision Avoidance Measures**
  Care must be taken to avoid potential collisions between moving components, and between moving and stationary machine parts when multiaxis toolpaths are being generated. As illustrated in Figure 9-4, this particular control focuses on means to avoid collisions, particularly between the cutter, arbor, holder, and the workpiece fixture assembly.

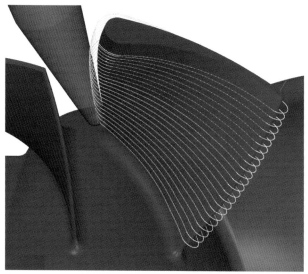

Figure 9-4 *Dynamic shank collision avoidance.*

- **Stock Recognition Roughing Strategies**
  Stock recognition during roughing will save time. Illustrated in Figure 9-5, stock recognition trims the toolpath to the stock size. This stock can be the initial CAD data or the in-process material created by previous milling operations. Multiaxis roughing can be a time-consuming affair and this feature is a must-have in creating efficient roughing toolpaths.

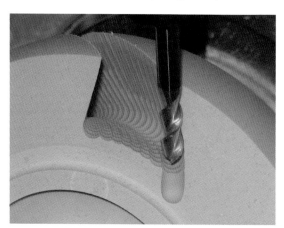

Figure 9-5 *Plunge roughing, using stock recognition.*

- **Machine Simulation**

  Machine simulation eliminates the guesswork and the need to prove-out new machining processes on real machines. Using a real machine to prove-out a toolpath wastes valuable production time and risks potential collisions. User-friendly and powerful virtual machine simulations, as shown in Figure 9-6, can improve productivity tremendously, but care must be taken to configure them properly for each machine. Please refer to Chapter 7 for detailed examples.

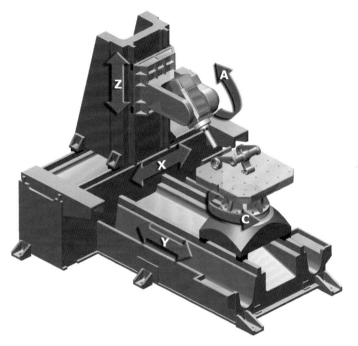

Figure 9-6 *Properly configured virtual 5-axis machines emulate the movements of real machines.*

- **Post Processor**

  A good post processor is the most important part of any multiaxis CAD/CAM software. Without post processing, parts can be cut only in the virtual world and not on real machines. The role of any CAM software is to generate code that will drive the movements of the axes on a CNC machine so that a part can be machined. The native CAM language must be translated to match with each machine's specific language. Customized multiaxis posts are usually an extra charge. It is important to find out if they are available for each specific machine and how much they will cost. A professional post processor is usually delivered with supporting documentation that explains its features and all the available switches to activate them. CAM software typically comes with a set of generic post processors, which are user-configurable. Ask if post-development training is available.

Multiaxis CAD/CAM Training

Because of the complexity of multiaxis machining, it is not recommended that multiaxis programmers be self-taught. Training is a very important part of getting the most out of the software purchase, and the best training is hands-on throughout the entire process. Training should include importing geometry, creating toolpaths, post processing, and simulating these toolpaths on a virtual machine. These steps represent half the job. The next step is to learn how to set up a real machine, find the machine's **Rotary Zero Point**, set the tool locations, load the toolpath into the machine's controller, and then cut the real part. Nothing can replace the feeling of excitement associated with running a new program on a real machine.

It is essential to find out if this kind of programming training is offered by the CAD/CAM company that responds to your request to quote. Many programmers take three- to five-day, canned training courses, which use pre-arranged training sessions and step-by-step instructions. It is possible to complete these training courses by simply following the carefully laid-out steps, but there is no requirement for the user to retain any insight into why they are following those steps. These user/trainees will get back to work and not know where to start. Very specific questions need to be asked about the training options offered.

On-line training courses are also an option. Some of these courses are very good, offering narrated videos, and hands-on training sessions. The appeal of these courses is that users can take them at home at their convenience.

CAD/CAM companies also offer on-site training. This arrangement ensures that the focus is on the operation and the parts for which the programs will be used. The danger with on-site training is interruptions. Care must be taken to stay on course at all times.

What happens after training? The chosen CAD/CAM company should provide applications support after training is complete. It is very helpful to have that support available as a safety net for at least the first few jobs.

How about update training? As mentioned earlier, CAD/CAM software is constantly evolving, and it is important to keep up with these changes by attending periodic update training sessions. User forums are also a very good way of keeping up with changes and a good way to exchange ideas with peers.

Behind the Scenes: CAD/CAM Software Development

The software that is ultimately chosen will have a profound effect on the business. Not only will the shop get the software functionality to run its machines, it will also be allying itself with a company that can provide years of experience and invaluable support. Considerable thought should be given to the company behind the software. A well-established, reputable, company can become a valuable asset and partner to the operation.

Understanding the development cycle of modern CAD/CAM systems can be helpful when software companies are being researched. The following behind-the-scenes look at the development cycle will illustrate why it is important to select a large, well-established company as opposed to a fly-by-night business.

CAD/CAM development is a very dynamic process. A successful CAD/CAM company consists of many teams of individuals working toward the same common goal. The individuals all strive to make powerful, flexible, and user-friendly software for the end-user. This task is difficult because the more adaptive and powerful the software is, the more complex it becomes. Complexity and ease-of-use often conflict with each other, and writers of good software strive to find a balance between the two.

Imagination is a very important and fundamental part of CAD/CAM development, but it can be tricky because it must be tempered with today's (and tomorrow's) hardware limitations. Theoretical possibilities are always restricted by current hardware limitations. CAD/CAM design is a long-term, ongoing project, and hardware advances must be correctly anticipated and implemented into the software.

Software development planning is done by mixed groups of individuals who include software engineers, mechanical engineers, applications engineers, sales, and marketing people. These groups are also heavily influenced by feedback from existing users. Existing users help these groups make up the "wish list" of new tools, as well as the recommended improvements slotted for the next software release. The software developers take a close look at the "wish list" and determine what can be done, when, and how.

Once the software development team has produced the first usable product, they will make it available to the rest of their teams, including quality control, applications, and post development. All these groups will conduct their own usability tests and provide feedback. The developers will use this feedback to fix bugs, improve the interaction, and make performance enhancements. This cycle is repeated continuously, until a stable, predictable, user-friendly Beta version of the software is created.

The Beta version is distributed to a special group of end-users who will conduct their own tests. At the same time, the software manufacturer's applications department will conduct more tests by cutting real parts on real machines.

Throughout this development process, everything is carefully documented. The technical documentation group writes Help files, and training manuals are developed and tested for each product.

At the very end of this planning, development, testing, and documentation process, a new version of the software is launched and monitored at every step. At that point, a dedicated technical support team is ready to assist customers with any issues that may arise.

But this point is not the end of the development process. The planning group keeps on dreaming and making new plans. The software development team stays busy working on those plans, and so on. A good software company has large teams of professionals in order to be able to continually develop new and improved software tools. The work is never done because it is literally on the leading edge of technology.

General Guidelines for Researching CAD/CAM Software

Start CAD/CAM research online. This approach can be a great way to compare the features and benefits of several different software packages. Many sites include demo video files, which can provide a good feel for the software's interface and will often illustrate the software's newest features. The web site will also indicate details of any local reseller in your area.

Conversations with peers or with companies with whom the shop will work are useful to learn what kind of software they are using and why. Ask people if they are happy with the local support, and was the software easy or hard to learn? Can files from outside sources be imported and exported easily? Were there any hidden costs? Is the local reseller reputable? Would they recommend the software they are using?

Visits to tradeshows are strongly recommended. Tradeshow demonstrations are short and are geared to show off the latest hot features of the software. Visiting software companies at tradeshows also provides the opportunity to talk directly to their corporate staff, and the staff can include people from all the different groups responsible for the software development. Chances are that the local reseller might also be on hand to explain specific features and services. Such visits are a prime opportunity to learn whether you would like working with the firm's employees, and to see if they are genuinely trying to help you or just trying to make a sale.

Most of today's modern CAD/CAM packages have very similar features, making it extremely difficult to compare them with each other. Another problem is that the packages are always subject to development, and therefore are constantly changing. Beware of anyone who makes comparisons between competing CAD/CAM systems, and beware even more of people who are trying to make a sale by putting others down. A tradeshow is a great opportunity to meet the people who develop and support the software. In addition to looking at the latest hot features of the software, take the time to assess the people you would be working with if you decided to purchase the software. Are they enthusiastic about their product? Are they behaving like a team, or are they shifty, disinterested and unhelpful?

The following are among important questions that should be asked when visiting software companies:

- Can you start small and then increase functionality as your business grows? Many software companies offer different levels of the software. Find out if you can buy only the functionality you need today, and add to it later as the business develops.

- Find out where your local reseller is located, and try to meet someone from the company. Ask questions regarding training, support, post processors, and other aspects of purchase. Make sure you are comfortable with the reseller because the support you receive can make or break your software experience.

- How established is the software manufacturer? It is a good idea to find a reputable company with a large user-base and support network. Find out how many programming seats are used worldwide. Is use of the particular software at which you are looking taught at trade schools or colleges? You may want to consider how easy or difficult it may be to find employees that already know how to operate your software of choice.

The next important step is to set up a demonstration at your plant. The local sales representative should visit your shop, look at your operation, and based on what kind of work you do, evaluate whether the software is the right fit for you. If it is, he/she can also recommend the proper software functionality you need. Beware of sales representatives who start with "Do I have a solution for you!" even before they see the type of work you do.

NOTES:

10

Putting It All Together

By now, readers should have a good grasp of the multiaxis machining process, with a clear understanding of the different types of machines, multiaxis toolpath types and machining techniques, multiaxis CAD/CAM controls, simulation options, and how they all fit together. To test your new knowledge, try to answer the following questions. Answering the questions successfully means that you are ready to break into the fast growing multiaxis machining world.

All questions will be answered on subsequent pages, and these answers can serve as a quick reference guide for the most important lessons learned in this book.

QUIZ

1.) Name three benefits to using multiaxis machining techniques.

1. _____

2. _____

3. _____

2.) Describe a standard 5-axis machine?

_____

_____

_____

3.) Which of the following is the standard axis convention?

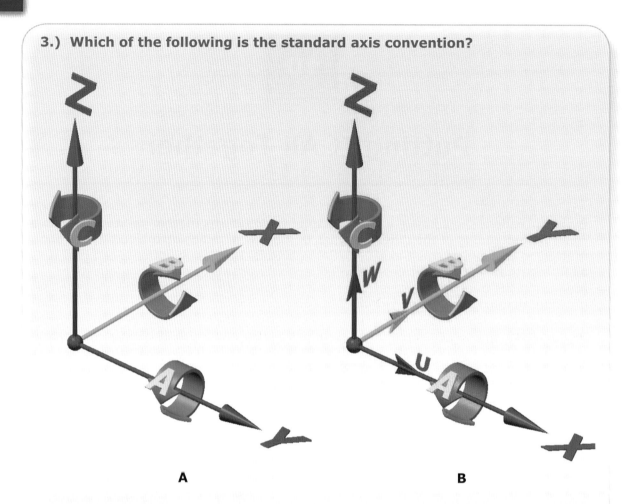

A   B

4.) What are the three major multiaxis machine types?

1. _____

2. _____

3. _____

5.) What are the three major building blocks of a CNC machine? *(Circle three.)*

- Machine table servo drive system
- Spindle RPM and horsepower
- Physical properties of the machine
- Chip conveyor unit

- CNC controller capabilities
- CNC drive system
- Linear table limit switches

6.) What are the most important physical positions of a multiaxis machine?

- Center of gravity, Home Base

- Program Home Base, Incremental Zero Position, Spindle type

- Machine Home Position, Machine Zero Position, Program Zero Position

7.) What tools are needed to find the Machine Rotary Zero Position (MRZP)? *(Circle two.)*

- level

- edge finder

- dial indicator

- maintenance manual

- hammer

8.) Describe indexing/rotary positioning work.

_____

_____

_____

_____

9.) What is a post processor?

_____

_____

_____

_____

10.) What is the definition of an axis?

_____

_____

_____

11.) Definition of a simultaneous 5-axis toolpath: All 5 axes of the machine tool must be continuously moving while cutting to be considered a simultaneous multiaxis toolpath.

- True

- False

12.) What are the three common simultaneous multiaxis CAM toolpath controls?

- cut pattern, tool axis control, tool tip control

- toolpath type, cutter control, cut pattern

- cut pattern, toolpath control, feedrate control

ANSWERS

1.) Why use multiaxis machining techniques?

- Multiaxis machining techniques are used to manufacture parts more efficiently and accurately by eliminating extra set-ups and fixturing.

- Standard shorter tooling can be used, which results in the ability to rough more aggressively, while increasing tool life.

- A more precise surface finish can be achieved by avoiding contact with the non-spinning dead center of the tool.

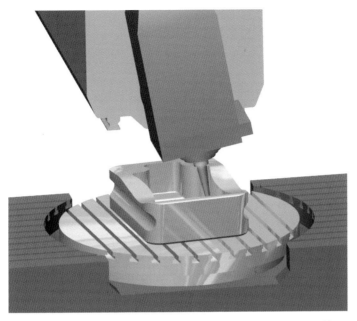

Figure 10-1 *Multiaxis machining manufactures parts more efficiently, increases tool life, and produces a more precise surface finish.*

2.) What is a standard 5-axis machine?

This is a trick question! There is no such thing as a standard 5-axis machine. Multiaxis machines are available in many shapes and forms. Figure 10-2 shows examples of the various types of 5-axis machines.

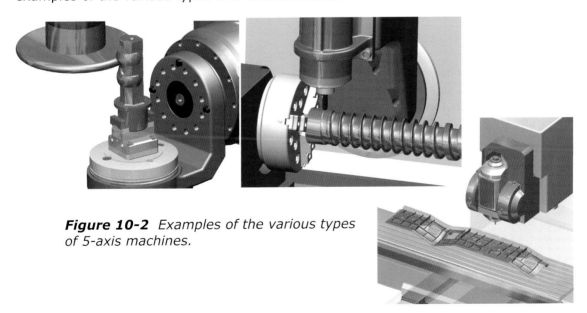

Figure 10-2 *Examples of the various types of 5-axis machines.*

3.) What is the standard axis convention?

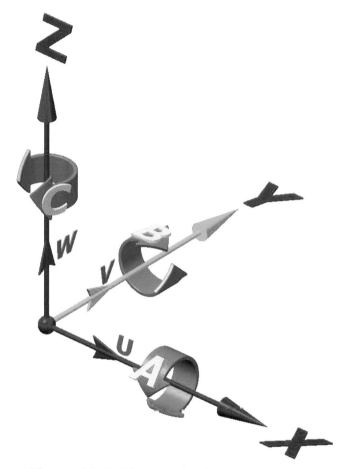

Figure 10-3 *The standard axis convention.*

The X, Y, Z linear axes shown in Figure 10-3, representing the Cartesian coordinate system, move in straight lines, in plus and minus directions. The A, B, and C rotary axes rotate about the X, Y, and Z axes respectively. The U, V, and W axes move in straight lines, parallel with the X, Y, and Z axes respectively.

4.) What are the three major multiaxis machine types?

> **TABLE/TABLE**
>
> **HEAD/TABLE**
>
> **HEAD/HEAD**

Table/Table Multiaxis Machines

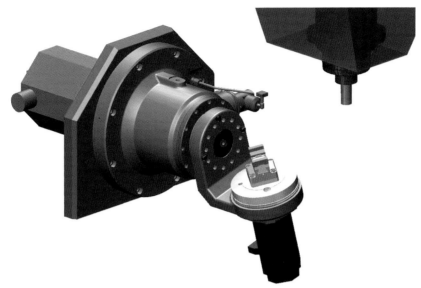

Figure 10-4 *Table/Table machines can be configured vertically or horizontally.*

Table/Table multiaxis machines can be configured vertically or horizontally, as shown in Figure 10-4. The rotary motions are executed by the dual rotary table of the machine. The rotary table carries another rotary table, which in turn carries the fixture and the part. With these machine types, the part is physically rotated around the tool. The weight of the part and fixture need to be handled by the machine's rotary devices, so inertia will be a factor when considering fast movements.

Head/Table Multiaxis Machines

Figure 10-5 *Head/Table machines are very capable and versatile.*

Head/Table machines are arguably the most capable of the three groups. They can machine large, heavy parts. On some **Head/Table** machines, the work piece is held by a rotary table and is supported by a tailstock, as shown in Figure 10-5. The work piece rotates around its own axis. The pivoting head only carries the weight of the tool and it handles the cutting pressures generated as it articulates around the work piece.

The rotary axis on these machines usually has unlimited rotary motion. Some can even spin the rotary as a lathe would. The secondary pivoting axis has an upper and lower rotary/pivoting limit.

Head/Head Multiaxis Machines

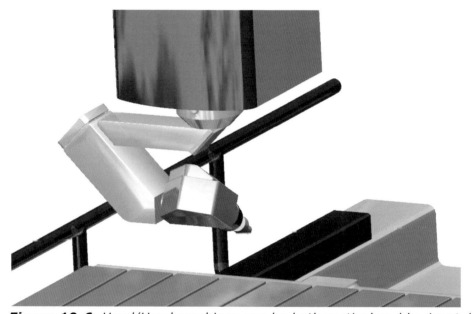

Figure 10-6 *Head/Head machines can be both vertical and horizontal.*

On **Head/Head** machines, an example of which us shown in Figure 10-6, all rotary/pivoting motions are executed by the head of the machine. Head/Head machines can be both vertical and horizontal, where one axis has limited motion. Some can change heads in addition to tools. Heads can be straight, 90 degree, nutating, or continuously articulating. In addition to milling, these machines can also be outfitted to manipulate a water-jet or a laser.

5.) What are the three major building blocks of a CNC machine?

1. The physical properties of the machine
The physical properties of the machine are represented by the machine's **skeleton**. Every machine is built on a unique base. The quality of the iron gives the machine its rigidity. The linear and rotary axes are stacked first onto the base, then onto each other. The quality of the linear slides and rotary bearings give the machine

its flexibility and potential accuracy. The spindle motor's torque and horsepower further define the character of the physical machine.

2. The CNC drive system
The CNC drive system represents the **muscles** of the machine. The CNC drive system consists of components designed to move the machine's linear and rotary axes. These components include the servo motors, drive system, and ball screws, which are responsible for moving the machine's linear and rotary components in a smooth, precise, and rapid manner.

3. CNC controller capabilities
The CNC controller is the **brain** of the machine. Data handling, available on-board memory size, and dynamic rotary synchronization controls, are some of the things controlled here.

6.) What are the most important physical positions of a multiaxis machine?

Machine Home Position - Most machinists recognize this position as the place to which all the axes move when the machine is initially turned on and Zero return is selected, as shown in Figure 10-7.

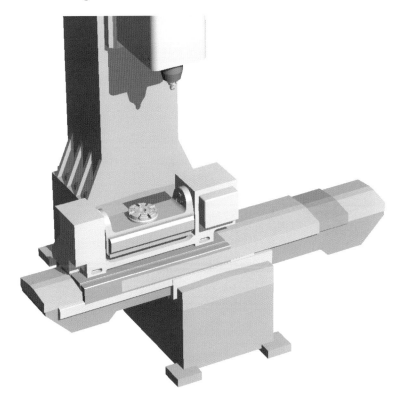

Figure 10-7 *Machine at Home Position.*

Machine Rotary Zero Position - Machine Zero Position is the intersection of the rotary/pivoting axes shown in Figure 10-8. This point may be unreachable by the machine.

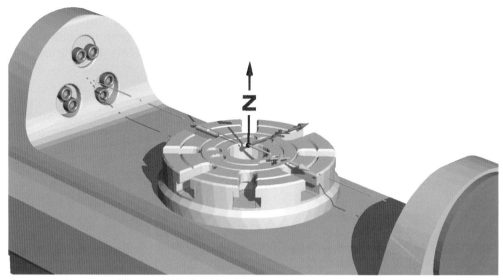

Figure 10-8 *Machine Rotary Zero Position.*

Program Zero Position - This position, shown in Figure 10-9, is also the part datum location in the CAM system.

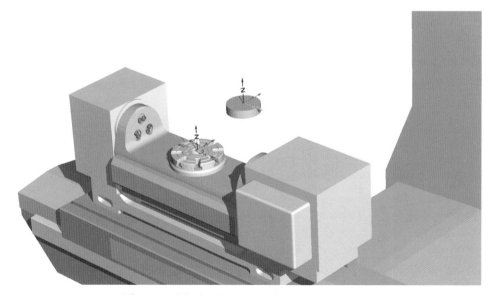

Figure 10-9 *Program Zero Position.*

7.) What tools are needed to find the Machine Rotary Zero Position (MRZP)?

The tools needed to find MRZP are a level and a dial indicator.

8.) Description of indexing/rotary positioning work

Most CAD/CAM systems let the user define multiple **Active Coordinate Systems** in space, and then create toolpaths using the orientation of each individual coordinate system. As shown in Figure 10-10, the Z-axes of these coordinate systems will align with the spindle, signaling the post processor to output rotary indexing commands into the NC code.

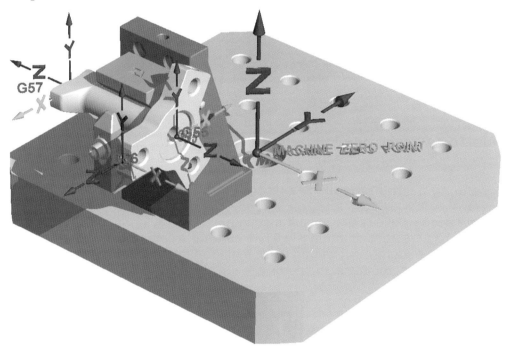

Figure 10-10 *Multiple Active Coordinate Systems.*

9.) What is a Post Processor?

CAD/CAM systems generate 5-axis vector lines along 3D paths. The 3D paths represent the tool motion as it follows the cut pattern. The vectors represent the tool axis direction (IJK vectors) as the tool follows the 3D (XYZ) pattern. Every vector represents a line of code. This information is written in a generic language.

The generic CAD/CAM code must be translated into a machine-readable language. This process is called post processing. A post processor will calculate motions needed on a specific machine to reproduce the CAM vector model, which will govern the machine's motions in order to cut the part. A different post processor is needed for every type of multiaxis machine.

10.) Definition of an axis

Any motion controlled by the NC controller, either linear and/or rotational is considered an axis.

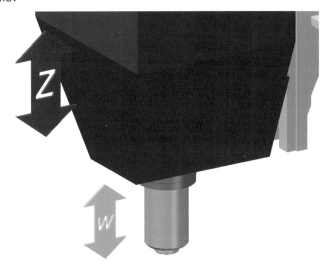

Figure 10-11 *In this example the spindle head and the quill move in the same direction, but are controlled by two separate commands, Z and W respectively.*

11.) Defining a simultaneous 5-axis toolpath

False. Most people believe that simultaneous multiaxis toolpaths must move all 5 axes of the machine tool continuously while cutting, when in fact a single rotary and linear combination is considered to be simultaneous multiaxis cutting motion. Typical simultaneous multiaxis toolpaths are illustrated in Figures 10-12 and 10-13.

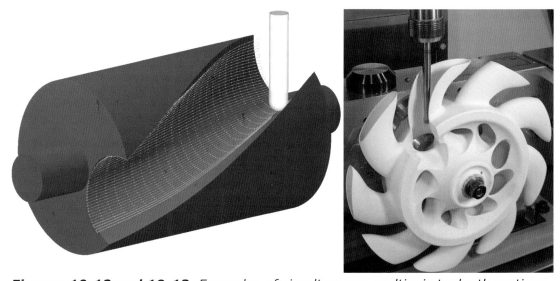

Figures 10-12 and 10-13 *Examples of simultaneous multiaxis toolpath motions.*

12.) What are the three common simultaneous multiaxis CAM toolpath controls?

1. **Cut Pattern** - Guides the tool along cutting directions.

2. **Tool Axis Control** - Controls the orientation of the tool's center axis as it follows the Cut Pattern.

3. **Tool Tip Control** - Controls the geometry to which the tool tip is compensated.

In addition to the above three major controls, quality CAD/CAM systems also offer additional collision control. Even near-miss collision avoidance of the cutter, shank, and holder can be checked against any part of the workpiece, fixture, or machine components.

Please refer to Chapter 6 for more detail.

More in Review: Multiaxis Machine Offsets

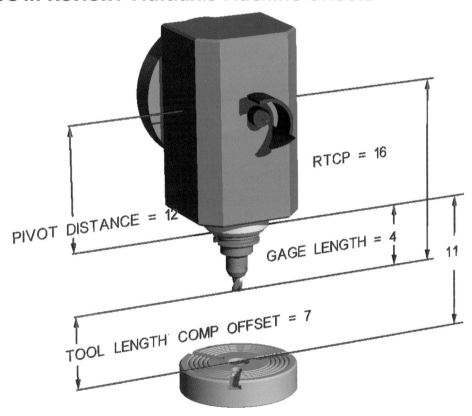

Figure 10-14 *In addition to Tool Length compensation, multiaxis machines use other offsets including Gage Length and Rotary Pivot Distance. The Rotary Tool Control Point Distance is the sum of Pivot Distance plus Gage Length Offset.*

Quick Reference: How to Find Machine Rotary Zero Position

For Table/Table Machines:

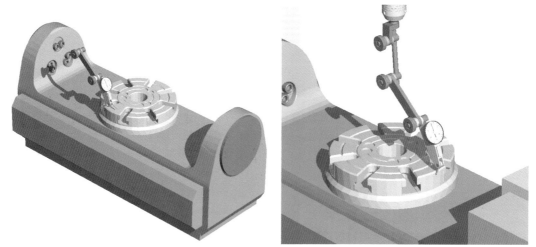

Figure 10-15 *Step 1: Level the A-axis.*   **Figure 10-16** *Step 2: Find X,Y center.*

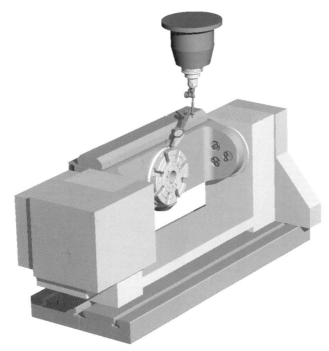

Figure 10-17 *Step 3: Rotate A+90 and set dial indicator to Zero.*

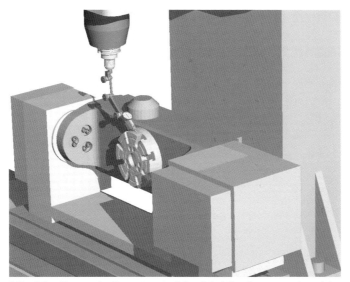

Figure 10-18 *Step 4: Rotate A-90. Dial indicator should read Zero*

Figure 10-19 *Step 5: Jog Z minus the radius of the rotary table diameter, and adjust gage tower height to match.*

Finding the Pivot Distance

For Head/Table and Head/Head Machines:

First, make sure that the machine head is in a perfect vertical orientation and that the spindle is running true.

Figure 10-20 Step 1: Use a dial indicator to check for vertical alignment.

Figure 10-21 Step2: Check if spindle is running true.

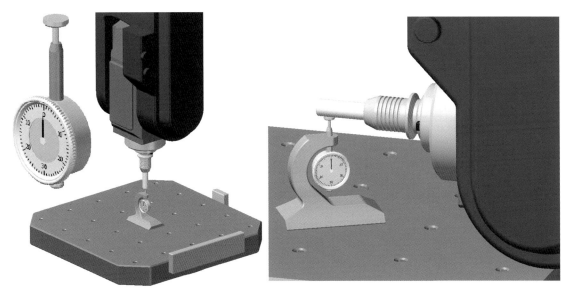

Figure 10-22 *Step 3: Record Z max.*   **Figure 10-23** *Step 4: Record Z min.*

Z max

Z min

GL - Gage Length

R - Dowel pin radius = .5000

Formula to calculate Pivot Distance:

PD = Z max − Z min − GL + R

Indexing/Rotary Positioning Work Overview

Also known as 3+2 machining, indexing/rotary positioning work, illustrated in Figure 10-24, is the most basic multiaxis concept. The rotary/pivoting axes are used only for positioning, and the cutting takes place with only the three linear axes moving. Indexing work is rigid and precise. It is recommended that all possible roughing operations be performed in this rigid state.

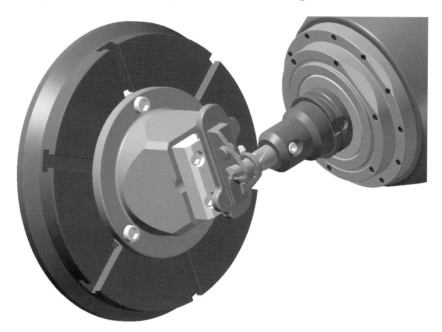

Figure 10-24 *Indexing/rotary positioning work is the most basic multiaxis concept.*

Picking a CAD/CAM System For Multiaxis Work

When selecting a CAD/CAM system for multiaxis work, make sure it is CAM oriented, and has a powerful CAD translator. The CAD translator is very important because it's likely that files will be received from many different sources. Make sure the CAM system has all the multiaxis controls, plus collision checking. Having an onboard, easy-to-use, machine simulation is a big plus, especially when project planning. Machine work envelope and machine component collision checking are required.

In addition to the above features, it is also very important to research the CAD/CAM system developer and the local dealer. Do they provide quality training and support, and do they have post processors for your machine?

Please refer to Chapter 9 for more detail.

Machine Simulation

Do not assume that machine simulation is used only for prove-outs with the sole aim of finding errors in the code. Instead, machine simulation should be regarded as an additional tool to help make clean, efficient, and accurate programs every time. Machine simulation permits testing of different approaches, different cutting strategies on different machines, without leaving the desk. There is also no need to tie down a machine for your prove-outs.

Machine simulation lets you build a replica virtual machine on the computer screen, where cutting processes can safely be simulated to make sure that the most effective cut has been created, that the part is located in the machine's "sweet spot," and that no fixtures, tools or any machine components are meeting unexpectedly.

In Conclusion

Congratulations on the commitment to become more informed about multiaxis machining! Multiaxis machining is a dynamic, constantly-evolving field, full of possibilities. Multiaxis machine tools will become more complex and capable, and CAD/CAM systems will develop additional capabilities to control them. Users will continually look for more capability, combined with ease of use, and this demand will pressure the machine builders and CAD/CAM developers to combine their efforts in building machine/controller combinations with built-in intelligence. As past trends show, these developments will open yet more possibilities, adding more complexity.

Creativity does not fit into a box, but knowing the basic concepts will allow engineers to think outside the box. Hopefully this book has demystified this field and inspired you to take the next step in training yourself to become more proficient and competitive with all the tools available. The best measure of competency in any field is mastery of the available tools. **Mere possession of more powerful tools doesn't make one more capable, but knowledge does.**

The manufacturing industry in general, and multiaxis machining in particular, is best suited for those who can think outside the box. There are always multiple ways to solve any problem and that solution always starts with oneself. The biggest secret of 5-axis machining is the realization that all the expensive CNC machinery, CAD/CAM, and simulation software are mere tools. Without the knowledge to use them properly, nothing can be accomplished. With the available tools and the right knowledge, all you have to do is imagine — by applying yourself, your imagination will become a reality.

NOTES:

Index

A

ABC linear axes, 15
Absolute coordinate system, 57
Accuracy, 9
Active coordinate systems, 25 - 27, 57, 59-61, 140
Actual part zero point, 27
Aligned universe, 62
Avoiding collisions, 45
Automatic tool changing, 16, 42-3
Axes, 3
Axis
   defined, 14
   substitution, 32

B

Ball-nose cutters, 10, 96, 130
Better surface finishes, 10

C

CAD/CAM systems, 3, 7, 27,
   capabilities, 139
   multiaxis considerations, 139
   origin, 60
   selecting, 137
   software development, 145
     researching, 146
   training, 144
Calculating pivot distance (PD), 33, 37-8, 169
CAM, multiaxis, 139
Cam-operated multiaxis machines, 3
Changeable spindle heads, 53
Checking positioning repeatability, 42
Circular
   interpolation, 73
   tolerance, 74
Clean core, 92
CNC
   controllers, 3, 76
capabilities, 13, 157
   drive systems, 13
Collision avoidance (see Avoiding collisions)
Common misconceptions, 4, 6, 7
Complexity of work, 120
Computer numerical control, 3, 92
Crashing, 117
Cut pattern, 79, 86-94, 140, 161

Cutting

   control, 97
   direction, 100
   strategies, 45, 70, 103, 117, 138, 167
   variable-pitch thread, 67

D

Dedicated multiaxis machines, 9, 10
Designations and directions of multiaxis machine
   movements, 15
Desired cutter area, engaging,10
Dovetail effect, 98
Dynamic
   control of tool axis, 90, 98
   rotary fixture offset, 16, 27-8, 36

E

Effective work envelope, 16
Engaging desired cutter area, 10
Extrusion milling machine, 123

F

Fanuc program, 34
Feedrate, 72
   dynamic changes, 138
   inverse time, 74-6
   optimization, 138
   standard time, 74
Finding the
   center of rotation, 21, 27-8
   pivot distance, 33, 36-9, 161, 164
   XY zero, 23
5-axis
   machine terms, 13
   vector lines, 76, 159
4-axis
   machines, 39
   positioning, 7
Fixtures, 47

G

Gage
   length (GL), 36-9, 161
   tower, 24, 163
Gantry type head/head machines, 122, 134
G-codes, 29, 30, 56, 104-106

simulation, 105
G-90 code, 29, 30
G-91 code, 29, 30
Graphical user interface, 116

H

Head/head multiaxis machines, 18, 36-7, 115-6, 121-2, 134, 156, 164
    bridge type, 122
    gantry type, 122, 134
    laser cutting machine, 116, 135
    water-jet milling machine, 116, 134
Head/table multiaxis machines, 18, 31, 36, 113-4, 123-4, 155
    aerospace, automotive applications, 129, 133
    milling engine head ports, 125
    milling long rotary parts, 124
    mold and die applications, 130
    nutating head combinations, 129
    rotary table, tilting head, 128-30
    various configurations, 124-9
    with long X-axis travel, 123
How CNC machines work, 56
History of 5-axis machining, 3

I

Indexing, 21, 44, 51, 55, 133
    fixtures, 51
    methods, 51
    toolpaths, 49
    with rotary devices, 52
    work, 49, 55
Industrial robots, 135
Interpolation
    circular, 73
    linear, 73
Inverse time feedrate, 72-4, 76

L

Laser scanners, 95
Lead and lag in milling, 100
Limitations, 46
Linear
    axis, 14-6, 34, 49, 74, 106, 121, 166
    interpolation, 73
Local coordinate systems, 25-7, 56-8, 61-2, 117

M

Machine
    active coordinate system, 25-7, 57-61, 140, 159
    building virtual, 64, 113-6, 116-7, 139, 143-4, 167
    business end, 64, 107, 125
    coordinate systems, 25-7, 56-7, 61-2, 140, 159
    home position, 16, 57, 60, 78, 157
    local coordinate systems, 25, 26, 61
    rotary
        center point, 60
        home position (MRHP), 17
        zero point, 21, 25-7, 36, 60-2, 71-2, 144
        zero position (MRZP), 16-7, 21, 25, 27, 36, 117, 158-9, 162
        simulation, 27, 63-4, 98, 103-6, 143, 165-7
        graphical user interfaces, 116
            using, 117
Machining
    center configuration, 108-110
    complex workpieces, 5
    engine components, 20
    profiling, 115
    program, 29
    routines, 5, 104, 138
    spiral bevel gears, 68
Machsim software, 106
Maintenance issues, 40
Manual data input (MDI), 25, 116
Master
    coordinate system, 60
    zero, 26
M-code, 21, 43, 60
Milling machines with five or more axes, 43
Modeling, 62-4, 71, 101, 111, 107-8, 116, 137-9, 159
Multiaxis machines, 3-6, 8, 17-9, 40, 74, 124, 153
    cam type, 3, 140
    dedicated, 6, 9-10, 21, 39, 52-3, 110, 120
    designations and directions, 15
    physical properties, 13, 156
    roughing, 21, 101, 130, 140-2, 166
Multiple nesting, 58, 61

N

Nesting positions, 25, 26, 56-8, 61
New possibilities, 11, 121
Numbers of parts, 120

Numerical control, 3

O

Old school simulation, 104
One zero method, 60
Optimum work envelope, 70
Origin, 26, 60

P

Pallet changers, 40, 54, 107-8
Part
    datum, 17, 21, 27, 58, 158
    zero point (PZP), 27-8
Plunge roughing, 101-2, 142
Point clouds, 95
Probes and probing, 94-5, 103-4
    sub-routines, 104
Physical properties of 5-axis machines, 13
Pivot
    distance, 33
    point, 37-9
Pivoting spindle heads, 18, 32-6, 38, 124, 156, 166
Pocket milling, 5, 86, 121, 137-9
Positioning work, 5, 7, 8, 13, 20-1, 26, 42, 49, 52, 159, 166
Post
    processing, 3, 4, 8, 34, 40, 76-8, 103-6, 138, 143-7, 159, 166
    processor, 3, 4, 8, 39, 40, 76-9, 104-6, 116, 138, 143, 147, 159
Probing routines, 104-5
Program
    manual editing, 104
    subroutines, 9, 43-4, 104
    zero position (PZP), 16-8, 25, 32, 117, 158-9, 162
Programming, 3, 5, 18, 26, 45-6, 56, 62, 71, 103, 105, 138, 144, 147
    considerations, 46
    languages, 3
    limitations, 46

Q

Questions and answers, 46, 144, 147, 149
    physical positions, 151, 157
    standard axis convention, 150, 154

R

Repeating patterns, 104
Rotary
    and pivoting axes, 32, 74
    axis, 16, 21, 33, 42, 60, 71, 74, 107, 121, 156
    devices, 16, 18, 21, 51-2, 116, 155
    indexing mechanisms, 5, 54
    mechanisms, 6, 19, 20, 39, 40-3, 52-3, 71
    tool control point (RTCP), 33-4, 36, 161
    zero positions, 16
Rotary tables, 5, 8, 9, 18, 24, 27-8, 31-2, 130-2, 155-6, 163
    brakes, 21, 40, 52, 104
    devices, 16, 18-9, 21, 51-2, 77, 109, 116, 126, 155
    dynamic fixture offset (RTDFO), 16, 27-8, 36
    single and dual, 6, 8, 18, 39, 119
Roughing, 11, 21, 101-2, 130, 140-2, 152, 166
Routines, 5, 40, 42, 104-5

S

Second rotary table, 18
Selecting machines, 119
Selecting software, 137
Simulation, 19, 27, 47, 63-4, 98, 103-17, 138, 166-7
Simultaneous
    cutting motions, 10, 71
    milling techniques, 21
    multiaxis toolpath controls, 79, 101, 152, 161
    toolpaths, 5, 48, 65, 78, 103, 105, 107, 121
Special-purpose software, 137
Spindle heads, changeable, 31, 53
Spiral splines, 99
Stacked errors, 9
Standard multiaxis nomenclature, 15
Stock (material) options, 47, 102
    recognition, 142
Sub-routines, 3, 43, 104
Surface finishes, better, 5,10
System
    origin, 60
    view, 27

T

Table/table multiaxis machines, 18-9, 24, 110, 125, 132, 155, 162
    with port-milling attachment, 125

trunnion and rock and roll fixtures, 71, 111, 132
   various applications, 133
3D surfacing toolpaths, 5
Tilting spindle heads, 31
Tombstone fixtures, 6, 40, 58-9, 108
Tool
   axis control, 79, 86, 89, 91-2, 98, 139, 141, 161,
   length offsets, 18, 24, 117
   lists, 46, 140, 145
   paths
     for lathes, 138
     simultaneous, 65
     plane with origin, 27
     tip control, 79, 90-91, 141
Tradeshows, 146
Training, 144
2 + 3 positioning, 49

U
Using motions XYZ and C, 67
Unlocked rotary drives, 11
UVW linear axes, 15

V
Vericut software, 1, 95, 106, 116-117
Verification system, 27, 104
Visiting software companies, 1, 146-7
Virtual machine, 103, 105
   building, 106
   components and models, 107
   configuring for simulation, 105
   kinematic component tree, 107
   skeleton, 106

W
Wire frames, 79, 103, 139-40
World zero, 26, 60

X
XYZ linear axes, 15, 32, 66-7, 74

Z
Zeroing the indicator, 22
Zero position, 17, 21, 117, 158, 162
Z-Maximum, 37
Z-Minimum, 38

Virtual Machining CD

All the images on this CD, including the virtual machines, were modeled using Mastercam® (CNC Software, Inc.). The virtual machines were brought to life using the machine simulation capabilities of MachSim (ModuleWorks) and VERICUT® (CGTech).

Installation

The enclosed CD should run automatically when inserted into a CD-ROM drive. If the auto-run feature does not work, please use File Manager to navigate to the CD. Find the file called Index.html and double-click it.

System Requirements

The CD was built to run optimally on a PC with:

- Windows XP or Vista
- Internet Explorer (Version 7) or higher
- 1024 x 768 resolution (or higher)
- Adobe® Acrobat Reader® installed. (Go to http://www.adobe.com/downloads/ to install a free version.)
- Apple QuickTime plug-in installed. (Go to http://www.apple.com/quicktime to install a free version.)
- If you install this CD on your hard disk, you will need 650 MB free space.

Virtual Machining CD Contents:

- **Over 25 Interactive Machine Simulations** - Self-extracting executable files launch interactive machine simulation sessions. Take control of all aspects of the simulation, including view manipulation, simulation speed, and individual axis control. Look at the machining process from various views impossible to see on a real machine. This offers a unique visualization to help understand a variety of multiaxis machining concepts.

- **Real Machining Videos** - Watch a real 5-axis machine perform several different multiaxis cutting routines on complex simultaneous 5-axis parts.

- **Virtual Machine Simulation Videos** - Watch VERICUT in action as it executes machine simulation and verification on over a half dozen different examples of complex multiaxis parts.

- **Printable PDF Files** - Quick Reference guides for the most important aspects of setting up a 5-axis machine and common multiaxis concepts all available as easy print-outs.

- **Image Gallery** - See full color examples of many of the parts and machines found throughout the book.

Technical Questions:

Please email your questions to info@industrialpress.com or to the author at karlo.apro@gmail.com. Or visit www.5axissecrets.wordpress.com and go to the link for FAQs.